Smart Users for Energy and Societal Transition

Smart Users for Energy and Societal Transition

Benoît Robyns
Claude Lenglet
Hervé Barry
Malik Bozzo-Rey

WILEY

First published 2023 in Great Britain and the United States by ISTE Ltd and John Wiley & Sons, Inc.

ISTE Ltd
27-37 St George's Road
London SW19 4EU
UK

www.iste.co.uk

John Wiley & Sons, Inc.
111 River Street
Hoboken, NJ 07030
USA

www.wiley.com

Library of Congress Control Number: 2022951689

British Library Cataloguing-in-Publication Data
A CIP record for this book is available from the British Library
ISBN 978-1-78630-735-4

Contents

Foreword by Pierre Giorgini

When it comes to climate issues and the changes imposed by its challenges, few comprehensive works exist. Combining technologies and social sciences is a feat in a university context, which tends to separate disciplines. We believe that this work, which is of high scientific rigor, is based above all on experience. We feel that it is this experience that lends itself to a transdisciplinary vision.

However, it is a common sentiment that when the disciplines converge, in part the right solutions can be found. Indeed, when creating representations and conceptions about the challenges to be met and the solutions to be implemented, we come across two extremes: on the one hand, a radical return to an energy sobriety imposed from above in the form of a dictatorship of the general interest, and on the other hand, a "techno-centric" headlong rush that places bets on future solutions, while allowing us to avoid fundamentally questioning our lifestyles, our consumption patterns and our relationship to happiness, which leads to the depletion of natural resources.

In this way, the path toward the common good arises as a third path. This path is enlightened and guided by this book. It becomes clear that we ought to invest and act in three directions that complement each other, but which are most importantly not exclusive. The first is the path to *reduce greenhouse gas emissions as much as possible* while maintaining a global vision of socioeconomic impact. The second concerns proactively *repairing systems wherever possible* (biodiversity, natural ecosystems), and the third is about *adaptation and limitation of the consequences of collapses*, made possible due to human intelligence.

In these three ways, technology will be called upon as it plays an important role, but a role which can be constantly reexamined within the framework of a reconfiguration of technosciences based on use cases, symbiosis, and integrating both ethical questions and those that concern the common good. The approach must be fully integrated, meaning that each designer will constantly have to question themselves with each design they create, asking questions like "What am I trying to solve?" "Why exactly does my solution not generate more external problems externalities than it solves?" "Is this well defined?" "Does it serve the long-term interests of those I'm designing for or other interests which are economic, geopolitical, etc. in nature?" and "What ethical blindspots might I have for the future?"

The hope is to "build on" and enable an ongoing epistemic metamorphosis. Here, we can refer to the concept of metamorphosis described by anthropologist Alain de Vulpian, as well as the emergence of endo-contributivity connection and homo holopticus[1], which dominate new ideas in all scientific, technical and social fields. These point to a hope that a new paradigm of cooperation will open us to a more desirable future. Endo-contributory requirements [GIO 21a], or co-elaboration, appear frequently. Systems can no longer be controlled from the outside because they are too complex. Intelligence must be distributed and embedded at the center of the components which will offer the agility and the capacity for a bifurcation that is necessary, because of a more flexible dynamic and the ability to simultaneously combine the local and the global (holopticism).

Alain de Vulpian, who passed away in 2021, worked for almost 70 years on the weak metamorphosis signals whose beginning he dated to a century ago in the West. He speaks of anthropological metamorphosis. A follower of Rogers, he identifies and analyzes through tens of thousands of interviews a fundamental evolution of human cognitive and social behavior. A follower of contemporary theses surrounding mankind, he suggests a striking analogy with the principles of functioning and evolution. For him, everything is alive in a holistic approach to reality. He applies the approaches of Francisco Varela and Ilya Prigogine to constantly refine his vision of this mutation which he describes as a humanist bifurcation. Then, in the recent discoveries of neuroscience, he confirms his analysis and discovers that there are tools to characterize this mutation from a cognitive point of view. He uses the concept of neural plasticity to suggest that cognitive bifurcation signals a new era relating to the relationship of the human brain to its environment.

---

1 Homo holopticus would come to designate the emergence of an "extended man" or "space man" in a network made up of humans, and the extension of the human through technology, the global architecture of which would dramatically accelerate its transition toward a holopticism that is integral, horizontal and vertical, temporal and spatial. The perception of the whole, of the global (produced by the entire network) and of its interactions with singular and local action is therefore what characterizes homo holopticus.

According to de Vulpian, a process of reversal has taken place because of the growing perception of the catastrophes engendered by an essentially technoscientific and rationalist vision of our relationship to nature. These are relationships with nature, other species and other humans which are dominated by competition, exploitation and the myth of progress, particularly technoscientific in nature. The progressive emergence of a global ecological consciousness is at the source of this new neural adaptation. Homo sapiens have relearnt to simultaneously mobilize the four dimensions of brain activity in the same way as Natufian hunter-gatherers did, a group which had a more holistic relationship with nature, other living beings and other humans which allowed them to survive. De Vulpian spoke of a "society-like-a-brain" using a single term, as well as a "citizen strategist". The strategic development of the community no longer involves permanent intermediary bodies. Hence, the crisis of trade unions, political parties and intermediary bodies in general.

In his two works [DEV 16, DEV 19], he defends the idea that the networking of socioperceptive brains develops a thin layer of "connected thinking" which can be brought out on a global scale, or a humanist noosphere (in reference to Teilhard de Chardin). Why humanist? Because it is similar to the phenomena of the living whose mechanisms have led to both constant hominization with the contingent purpose of survival, as well as to a constant process of humanization by a life force.

But, as with the evolution of living things, this evolution is not linear in nature. It can lead to phases of regression and of resistance from the powers that be, like those we are witnessing today. Indeed, this bifurcation disturbs all of our conceptions as well as all of our traditional hierarchies. It frightens many, appearing as a threat. This leads one's creative, empathetic and intuitive capacities to narrow, as well as to the phenomenon of humanity searching for a set of new certainties which are simple, radical and non-complex. According to de Vulpian, threat and fear lead to the activation of the retrograde brain, that of fight, flight and anger.

But, according to de Vulpian, this could be temporary if everyone took care to cultivate a level of humanist metamorphosis, to monitor it and above all to effectively educate young people about the four dimensions of their brain's capacities: spiritual, emotional-relational, sensorial and rational.

For this, it is necessary to encourage the current emergence of hybrid, heterarchic collectives[2]: to open and network self-organized communities, to promote learning about self-regulation in hybrid communities and encourage them to find a sense of balance. Thus, without sharp regression, a new type of rationality

---

2 Heterarchy is an organizational system which differs from hierachy because it promotes interrelation and cooperation between members rather than a bottom-up structure (Wikipedia).

may emerge, which is in any case part of our process of long-term cognitive evolution. Because it is, according to de Vulpian, the structure that allowed mankind to thrive in its sense of adventure. This book, which should be read without delay, masterfully invites us to do just this.

Pierre GIORGINI
Essayist, associate researcher in technology ethics
ETHICS laboratory, Université Catholique de Lille

Foreword by Xavier Bertrand

Ten years have passed since the rev3 dynamic was implemented in Hauts-de-France. It has also been 10 years since the Université Catholique de Lille launched its energy and societal transition program called "Live TREE" (Lille Vauban-Esquermes for energy, ecological and economic transition). The advent of these two concurrent anniversaries is not by chance. Rev3 was able to serve as a general framework, both "inspiration" and integration, for Live TREE, while Live TREE became from the outset one of the most ambitious and more interesting projects linked to rev3. Both approaches address the same issues, they share the same major objectives and they adhere to the same principles of action.

If the collective work *Smart Users for Energy and Societal Transition* largely describes the impressive initiative Live TREE, what can be said about rev3?

Rev3 is first and foremost the name given to a model, that of the "Third Industrial Revolution" (TIR), which was created by the American economist Jeremy Rifkin. According to Rifkin, industrial revolutions arise when a type of energy and a type of communication are articulated while in the process of domination. The First Industrial Revolution, which significantly marked our beautiful region, was triggered by the combined emergence of coal and railways. The Second Industrial Revolution came about due to the rise of oil and large networks (roads, electricity, fixed telephone infrastructure, etc.). The Third Industrial Revolution, which is now taking place, is marked by the emergence of renewable energies and the Internet. A prelude to a new development, it also aims to provide a major response to the major threat of climate change by describing the paths toward a low-carbon society. Rifkin's model is built on five pillars: development of renewable energies, buildings that "produce" electricity, energy storage solutions, smart energy networks and sustainable mobility, based in particular on electromobility. To a degree, we highlight these pillars in the work that follows. They are also present, to varying degrees, in the Live TREE program.

But in Hauts-de-France, rev3 goes further than Rifkin's original model. Due to the desire to apply the TIR to the regional territory, the two co-pilots, the regional council and the regional CCI (Chamber of Commerce and Industry), immediately wished to add a significant economic component to the "five pillars" previously stated, including more specifically the circular economy and the functional economy. When enhanced in this way, rev3 constitutes a development approach that is capable of responding to three major challenges that our societies face:

– energy transition, by promoting renewable energies within an effective energy mix which includes nuclear energy (an energy that is also carbon free);

– technological transition, by promoting research and innovation, thus ensuring that new products and new industrial processes are created;

– economic transition, with the establishment of new activities and the creation of new jobs.

Since 2013, rev3 has also become a real human adventure. It organizes a group of actors, which is quite remarkable, particularly in light of the multiplicity of projects in flight, the diversity of the partners involved and the durability of the undertaking, which we can claim without doubt. Rev3 is an illustration that is as concrete as the installation of three gigafactories of batteries, a technological platform dedicated to improving the energy efficiency of motors, the creation of a new financial tool (e.g. the "rev3 booklet"), student initiatives in the field of energy transition, etc. The projects are vast: it is estimated that there have been more than 1,500 of them between 2013 and 2022.

In fact, rev3 allows for the combined efforts of three distinct "worlds":

– first, the world of companies and their representatives. The company is the main player in rev3. They are at the origin of new activities and they create new jobs;

– next, the world of local authorities. It encourages initiatives and supports them with the appropriate technical and/or financial support;

– finally, the world of training and research.

We will reflect a little longer on the third category of actors, insofar as the work that follows highlights it with more attention.

Research and training are essential rev3 levers. Innovation is at the heart of rev3, whether technical or "societal" in nature (innovation in behavior, in organizations, in procedures, etc.). Therefore, research, both public and private, is to be mobilized so that flows of innovation can be generated which are able to maintain the dynamic. This should occur in connection with companies and within the framework of innovative ecosystems. The role of training is also crucial. In 2017, it was estimated that 85% of jobs in 2030 would not exist in 2017. Thus, it follows that the issue of new skills is particularly acute and requires implementing the various forms of training: initial, professional and, of course, higher, for the most qualified jobs. Universities and schools are therefore key players in the process.

This is why in 2018 Philippe Vasseur, former president of the Rev3 Mission, took the initiative to create a network of actors working in higher education and research on topics surrounding rev3. The name of this network is Unirev3, which has 33 members who signed an agreement that rallies them together around common ambitions and projects. It is important to note that the seven universities located in Hauts-de-France are part of this network, thus demonstrating their intent and their mobilization in favor of maintaining regional dynamics.

For its part, the Université Catholique de Lille has shown itself to be at the forefront of commitments and achievements, and in this respect, the Live TREE program is quite exemplary. First, in terms of actions, Live TREE uses the "fundamentals" created by rev3: that is to say, the reduction of energy consumption, the development of the Internet of energy, the production of renewable energy, the storage of electrical energy, the soft mobility, etc. In this case, the various projects are applied to the building stock of the university, in particular the "Rizomm" building, which really exemplifies rev3. Second, in the same way that rev3 is not limited to a "technicalist" approach, but rather incorporates a strong societal component, Live TREE is especially open to human and social sciences and thus seeks to develop an interdisciplinary approach in the most effective way through the design and implementation of operations. The ethical dimension of design is also constantly questioned. Finally, and this is also a characteristic that complements rev3, Live TREE attaches considerable importance to the partnership of actors for the various projects carried out: not only between internal university actors (students, teachers, researchers, administrative staff and university technicians), but also in connection with the inhabitants of the district, businesses and local authorities. This of course includes the region which is associated with the program and which supports it financially.

And now, how ought we to move forward? Faced with the magnitude of challenges ahead, we cannot be satisfied with the status quo. Moreover, in recent months which have been marked by a health crisis and uncertain geopolitical contexts, particularly important issues have been highlighted: namely, the resilience of our territories, and the sovereignty of industry and energy. However, rev3 is a promising way to respond to all of these concerns. We therefore need to reinforce and accelerate rev3. The region wants to contribute as effectively as possible and with all means at its disposal to the necessary transitions: economic, ecological and societal. Rev3 has already permeated regional policies to a large extent. My wish is that rev3 becomes a sort of backbone, to become a real "marker" for the regional mandate which began in 2021. Frédéric Motte, who is the new president of Mission rev3, has been working hard in collaboration with the regional services and our partners, in particular the regional CCI, to implement these new resolutions. A 2022–2027 "roadmap" has been submitted for the approval of regional advisers. It specifies the projects to prioritize in terms of sectors: energy mix, decarbonization, sustainable building, sustainable mobility, agriculture/bioeconomy and circular economy. It provides for an even wider deployment of rev3 in Hauts-de-France. It aims to strengthen relations between companies and universities and schools, whether on research/innovation or training topics. Finally, this new roadmap intends to prioritize developing the "citizenship" aspect of rev3, definitely not to give it additional evidence, but because the scale of the transformations to be carried out does indeed mean that everybody involved needs to mobilize. To achieve this, broad awareness-raising through various channels while reaching various audiences is essential. Students and the university community constitute one of these audiences. All in all, we believe that this book, *Smart Users for Energy and Societal Transition*, makes a significant contribution to current reflections, to the dissemination of ideas and, all in all, to progress toward our ambitions.

Xavier BERTRAND
President of the Hauts-de-France region

Introduction

The world and planet Earth are experiencing a serious ecological crisis, brought about by unprecedented energy and material consumption by humanity, which impact the climate and biodiversity in an irreversible way. This evolution, which is increasingly impacting the living conditions of humans, can be slowed down in order to allow for a more controlled energy and societal transition. The key is to adapt the consumption patterns of humans and their habitats so that less energy and materials extracted from the earth are consumed, and so that consumption is undertaken more intelligently.

The consumption of fossil fuels since the 18th century has undeniably allowed industry to develop, along with transport and standards of living, particularly in more industrialized countries. It has also contributed to an increase in greenhouse gases in the atmosphere, causing global warming. This phenomenon continues; if emissions of these gases are not quickly and drastically reduced, global warming will negatively impact the planet and our lifestyles, which will become increasingly unavoidable and increasingly difficult to live with. Though fossil fuels are the main source of CO_2 emissions, they are not the only one.

Energy and societal issues are linked, with energy occupying a very important place in our lifestyles. We may consider energy that is directly consumed (for heating, lighting, food, transport, running all electrical, digital and other devices), or energy needed to make the products we consume (to extract and transform materials, for crops, livestock, etc.), but also the impact of our lifestyles, which for example cause deforestation which reduces natural sinks of CO_2, which then lead to it being stored in the atmosphere.

The buildings in which we live, work, make our purchases or even those intended for leisure and sport are primary CO_2 emitters. The technologies used to build and operate them play an important role in how significant carbon emission levels are. However, the way we live with them and how we use them also contributes to these emissions. Buildings which are equipped with devices can better control energy. They can produce energy and store it, because of the new materials used to build these buildings but also increasingly because of new energy and digital technologies which work while ensuring that their occupants remain comfortable. Buildings therefore become smart building. However, technology alone is not enough to optimize the operation of a building and reduce its carbon footprint. Those who use a building also play an important role, due to their activities and their need for comfort, which vary from one individual to another. The question then arises as to whether in order for smart buildings to achieve their low carbon footprint objectives, they should not be used or inhabited by smart users[1], who are adapted or integrated into the building's intelligence. This provocative question will be one of the topics discussed in this book.

In this way, buildings are set to become intelligent, as are energy networks (smart grids), due to increased user involvement. These buildings interact with energy networks, integrating new practices of self-production and self-consumption of energy. All of these systems are interconnected by information systems which generate a technological convergence of energy between smart buildings, smart grids, the Internet of things and people.

Buildings and more broadly habitats associated with energy networks which are becoming more respectful of the environment and lower in carbon emissions constitute the foundations of a city in the future. If we associate quality of life with the return of nature to the city (or even urban agriculture), low-carbon transport such as public transport (transport is another major emitter of CO_2), spaces and organizations promoting living together, democratization of the means of information allowing inhabitants to be partners of the city and no longer only consumers, we will be able to formulate a smart city. The evolution of cities is important, because if cities today occupy 2% of the surface of the globe, they are home to 50% of the world's population, consume 75% of the energy produced and generate 80% of CO_2 emissions[2] [ROB 19]. The challenge for cities to become more sustainable, carbon neutral and resilient is therefore considerable.

---

1 According to the *Le Robert* dictionary, an intelligent person is someone who has the ability to know and understand, who is to a variable degree endowed with intelligence. According to Wikipedia, intelligence is the set of processes found in systems with varying levels of complexity, which are either living or not, which make it possible to understand, learn or adapt to new situations. In English, smart means intelligent, shrewd (*Cambridge* dictionary).

2 Available at: www.smartgrid-cre.fr.

Smart buildings, smart grids, smart cities – Are these concepts that implicate many promising technologies in the fight against climate change, especially if they are inhabited and used by smart users? We could optimize everything with appropriate algorithms. Why do not we use artificial intelligence, which with a lot of data (big data) [GIO 21a] could control all of the smart components of the system and thus help us to achieve optimal behaviors (of materials and humans) to save the planet? This is a "Big Brother" dream that actors in technosciences could seek to solidify soon. If technological solutions will form part of the solutions of the future, it is becoming increasingly clear that these solutions will not be optimal, or even that they may generate negative effects for the planet. This will be particularly true if they are not accepted by populations who appropriate them, because when technologies are desirable and economically viable, they can be transmitted and enriched with their creativity. Human rationality does not relate to the machine; individual and social acceptability are not automatic [ROB 19]. To the extent that the energy and societal transition becomes more than necessary to ensure decent living conditions for our children and grandchildren (current generations have a responsibility toward future generations that do not yet exist), the question of influencing lifestyles through information, new standards, the encouragement of new modes of consumption (energy and otherwise), coercive methods (laws and penalties) or various techniques of more subtle influences pose new ethical questions. We find ourselves at a crossroads between technosciences and societal issues.

Therefore, to succeed in the energy and societal transition, different scientific disciplines must work together, in order to develop interdisciplinary approaches between human and social sciences and engineering sciences, which are essential for a transition to be successful for all.

This work, after reviewing the necessary transition and the scenarios and possible solutions to mitigate the impact of climate change, will address:

– the development of ecosystems in cities, territories and universities to experiment with new solutions that enable energy and societal transition;

– elements that constitute future smart cities, with smart buildings and smart grids, by opening up interdisciplinary research perspectives;

– sociological questions related to the role of users in smart buildings, ranging from acceptance to involvement, assuming the implication of technologies and an environment that can be appropriated by users;

– the ethics of energy and societal transition or influencing behavior so that everyone works together to save the planet.

These subjects are covered in this book by an interdisciplinary team of authors: two engineers (an electrician, a building expert), a sociologist and a philosopher-ethicist.

Chapter 1 presents the connection between energy and societal issues. It allows us to set the context and explore the challenges of the energy and societal transition. Some perspectives which relate to climate change are identified, from denial and inaction to sustainable development expressed in particular by the 17 objectives identified by the UN, which pass through technosciences and the economy. One highlights a series of scenarios which can be imagined for the next 30 years to limit global warming by aiming for carbon neutrality by 2050 (assuming that carbon emitted by human activities is completely absorbed by plants, soil and the seas, or by technologies for capturing and storing the CO_2 that are being developed). These stories offer many solutions that have varying degrees of technological maturity and social acceptability, such as sobriety. Conditions inherent to the success of these scenarios and obstacles to their deployment are also identified. In short, a discussion is initiated which will be continued throughout this work.

Chapter 2 presents examples of cities and territories that have embarked on proactive energy transition by not hesitating to experiment with new technologies and lifestyles, which highlight a few key results. These examples are Copenhagen, Manchester, the Swiss project of the 2000 watt society and the Third Industrial Revolution in the Hauts-de-France region. Bringing these approaches together makes it possible to learn many lessons and provide some concrete answers to the questions raised considering the urgency of the environmental transition. In particular, it makes it possible to identify the "foundations bricks" and the cross-functionalities that are necessary for any energy and societal transition process.

Universities are ideal places for experimentation at scale and in real conditions of different axes of energy and societal transition. This is due to the support of interdisciplinary research, but also because of information, awareness and involvement of students and staff, as well as the populations within the territories in which they are anchored, through training and education missions. Chapter 3 gives some examples of universities around the world that are specifically targeting carbon neutrality and that aim to reach UN Sustainable Development Goals (UN SDGs). These examples include the universities in Manchester, Stockholm, Boston, Reading, and the University of British Columbia in Vancouver. Since 2014, in the Hauts-de-France region in the wake of the Third Industrial Revolution, the Université Catholique de Lille has been engaged in the "Live TREE" program (Lille Vauban-Esquermes in energy, ecological and economic transition). This program aims to support university establishments to achieve carbon neutrality, to develop a living laboratory on a real scale (via a living lab) as well as other challenges. This is achieved via experimental buildings in a sustainable and desirable district, which enable one to experiment and implement green mobility strategies, bring nature back to the city, develop the student experience by encouraging the active involvement of students in the transition, and develop transdisciplinary research between human and social sciences and engineering sciences, etc.

Energy is a fundamental issue when thinking about the transition. This is why energy networks are set to evolve strongly toward smart grids, just as buildings are set to evolve toward smart buildings. Chapter 4 introduces the concept of smart buildings as the nodes of smart grids. The challenge that comes from this aims to position the users, operators and owners of buildings at the heart of the approach, through modeling and dynamic supervision of buildings and blocks of mixed tertiary and residential buildings, integrating use cases and actors, with a view to transform them into intelligent nodes of a smart grid. This concept is a step toward developing smart cities, raising research questions (associating energy, buildings, transport, urban farms, digital technologies, citizen participation, etc.) which will be discussed in this chapter.

Chapter 5 asks a fundamental question about the buildings of the future: is an economical smart building without the cooperation of the occupants at all possible? The name smart building refers to the concrete modification of the technical contents of buildings, and therefore the way they function from a socio-technical perspective, using a very technical logic. Under these conditions, the appropriate designation of smart users, which was introduced previously, appears ambiguous, even paradoxical. This is because if the tendency inherent in the model is to ask nothing of the occupants themselves, what intelligence are we talking about, considering that occupants are never passive when they produce actions in accordance to their needs in a given moment? The paradox between the technicality of smart buildings and an impossible neutrality of the occupants is at the heart of the socio-technical problem of this new construction model, and so it should be questioned. Having said this, the paradox is ultimately very long-standing. The advent of smart buildings only serves to reconfigure the difficulty that those who design construction models face when integrating the parameters of use cases. Does the model of smart buildings more effectively overcome this genre of difficulty? This is the central socio-technical issue addressed in this chapter. The analysis developed in this chapter leverages the feedback that came from two buildings at the Université Catholique de Lille which were renovated and transformed into examples of smart building as part of the Live TREE program: the Rizomm building devoted to the faculties, and the HEI building of Junia, graduate school of science and engineering. Both share the same academic vocation, but radically differ when it comes to their socio-technical philosophies. These buildings have different ages: the oldest, HEI, dates back to 1885. Transforming such old buildings into smart buildings is thus obviously a challenge.

Chapter 6 discusses the ethical challenges inherent in the energy and societal transition. Climate issues and the necessary transition raise immense challenges such as the conflict between the interests of people living currently and those who will live in the future, as well as the prospect of a less prosperous future and/or with lives that are different from those lived by people today. This forces us to reassess what we understand by a "good life" or a "full life" (or a "more sober life"). Finally, if the basic needs of the majority of the world's population cannot be met in the future, then we are bound to face tragic situations where the choice between life and death will no longer simply be hypothetical. In other words, thinking about the ethics of the energy and societal transition involves rethinking and questioning three assumptions that have structured modern ethical thinking: first, the interests of present generations coincide with those of future generations, future people will be better off than us and favorable living conditions will continue indefinitely into the future. This chapter also deals with the way in which we can influence the behavior of individuals, in particular through public policies that target the greatest number of people. These have experienced a major turning point in recent years following the development of nudges (gentle encouragement given to an individual to modify their behavior without constraining or hindering them) and the integration of behavioral sciences in their creation. Finally, once the means have been identified, the question of their ethics and legitimacy can be raised. It will be shown that the energy and societal transition requires the unification of private ethics (individual) and public ethics (institutional). It presents a major modern challenge for our democracies: combining ethics and politics with respect to individuals while protecting their interests and those of future generations.

The issues are such that our way of seeing the world (the planet, humanity and biodiversity) and our relationship to the world must be transformed. Imaginations have to be modified, so that sources of happiness other than those that have a negative impact on the world are made more attractive, and sources of joy are sought out and rediscovered [GIO 20].

1

The Necessary Transition of the 21st Century

1.1. Introduction

The consumption of fossil fuels since the 18th century has undeniably enabled the development of industry, enhanced transport and increased the standard of living, particularly in the most industrialized countries, but it has also contributed to an increase in greenhouse gases in the atmosphere, provoking global warming in our planet. This phenomenon continues to worsen, so much so that if these gas emissions are not quickly and drastically reduced, global warming will impact the planet and our lifestyles even more, an issue that will become increasingly significant and difficult to live with. Though fossil fuels are the main source of CO_2 emissions, they are not the only one.

The first section of this chapter presents the connection between energy and societal issues. This will allow us to set the context, present the issues and bring out some initial aspects for reflection.

The second section presents some perspectives relating to climate change, from denial and inaction, to sustainable development, which includes technosciences and economics.

Finally, the last section brings out some scenarios that can be imagined for the next 30 years with the aim of limiting global warming. These scenarios offer many solutions that have varying levels of technological maturity and social acceptability. We will also identify a few conditions for these scenarios to be successful, as well as some obstacles to their deployment. A discussion will then be initiated which will continue throughout this book. The issue is so serious that our way of seeing the world (the planet, humanity and biodiversity) and our relationship to the world must be transformed. The way we imagine the world has to be modified, and other sources of happiness other than those that have a negative impact on the world are to be promoted, by seeking or rediscovering sources of joy [GIO 20].

1.2. Connection of energy and social issues

1.2.1. *Living energy*

It is undeniable that energy has enabled a tremendous amount of development within human societies, particularly in terms of food, health, education, technology, mobility, etc.

The use of energy that comes from nature – such as fire for food and heating, wind for mobility through turbines or water to drive motors – has always brought about a considerable pace of evolution within the societies that it has impacted.

Humanity has also used animals as a driving force, such as horses, as well as humans. However, the hydraulic energy brought about by water mills, which appeared in Antiquity, caused a reduction in slavery. By milling up to 150 kg of wheat per hour, a single mill cut down the work of about 40 slaves [BAR 14]. However, a decisive leap was to be made by humanity when it started exploiting fossil fuels in combination with ever more sophisticated technologies. According to the historian Jean-François Mouhot, the emergence of steam was most likely a necessary condition for the abolition of slavery. This is because the exploitation of fossil fuels led to an energy transition that made slave labor appear more superfluous. This has led to machines all but replacing forced labor in modern societies [MOU 11]. Reflecting on our current conditions, supplying the equivalent of only 4 € of daily food to a slave who supplies 100 W for 10 h of work, which is exhausting for any human being, or 1 kWh every day at the price of 4 €/kWh, means that the kWh price is 27 times higher than the 2019 value of public electricity in France [CAS 20]. The average French citizen currently contains the energy potential equivalent to 500 humans!

1.2.2. *Fossil fuel, deforestation, cattle rearing and climate*

For hundreds of thousands of years, the concentration of carbon dioxide (CO_2) in the Earth's atmosphere remained stable due to a balanced carbon cycle: the CO_2 emitted was essentially equivalent to the CO_2 absorbed, which leads to carbon neutrality. People are increasingly practicing deforestation and fossil combustion, that is to say, when what is emitted exceeds what is being absorbed, and the concentration of CO_2 in the Earth's atmosphere is increasing [GRA 16].

The increase in the greenhouse effect makes the global surface temperature of the planet rise. However, due to human activity, the concentration of greenhouse gases has exploded since the pre-industrial period (1750–1800). The concentration of CO_2, which is the main greenhouse gas, has increased by more than 30% since the pre-industrial era. The combined effects of all greenhouse gases (CO_2, methane, ozone, etc.) that we are seeing today amount to an increase of more than 50% in CO_2 since this period [ROB 21].

Between 1860 and 2010, the Earth's average surface temperature increased by 0.6°C, and 1.2°C if we extend the period until 2022. Different future scenarios predict that by 2100, if current energy sectors and consumption habits are not modified, we can expect temperatures to increase by another 1.5 to 7°C. This considerable increase would be accompanied, more specifically, by a rise in sea level of 20 cm to 1 m. Though the evolution of the climate appears to be irreversible, it is however possible to slow down this evolution by significantly reducing greenhouse gas emissions.

Natural CO_2 sinks such as soil, trees and oceans would only be able to absorb a little less than half of the CO_2 produced by humans (produced in 2000). In order to stabilize the concentration of CO_2 at its current level, it would therefore be necessary to immediately reduce emissions of this gas by 50%–70%. Even though this immediate reduction is seemingly impossible, it is urgent to act, because we are faced with a cumulative problem. Considering that the lifetime of carbon dioxide in the atmosphere is around a century, it will take several generations to make CO_2 levels stabilize to an acceptable level. In 2018, the Intergovernmental Panel on Climate Change (IPCC) estimated that in order to limit global warming to 1.5°C in 2100, CO_2 emissions would have to decrease by 45% in 2030 compared to 2010, and by 91% in 2050 [INT 18].

CO_2 is produced when all fossil fuels are combusted: oil, gas and coal. CO_2 emissions are about twice as high for coal as for natural gas, those related to oil situated between the two [ROB 21].

Global warming, and more broadly human activities, have an impact on biodiversity. Rates of extinction of species are predicted to be 50–560 times higher than when compared with those of a stable biodiversity. Here, we are talking about the sixth extinction. In addition to this, 11% of greenhouse gas emissions are due to deforestation and change relating to land use, as we are seeing uses that store less carbon such as producing palm oil [GRA 16].

Livestock is also a major emitter of greenhouse gases, constituting around 9% of CO_2 emissions worldwide. These emissions can mainly be attributed to the production and transportation of food; a production use case that requires agricultural land which as a consequence contributes to deforestation. The second source of emissions is the gastric fermentation of roaming animals. According to the IPCC, beef farming emits five times more CO_2 globally than pig or chicken farming.

Figure 1.1 highlights how CO_2 concentrations have increased (in parts per million, or ppm) in the atmosphere since 1700, with a dotted line which projects up to the year 2100, according to the worst-case scenario which has been provided by the IPCC scenario 8.5. This scenario implies business as usual, where a lack of intentional action to reduce CO_2 emissions leads to CO_2 concentrations above 700 ppm in 2100, compared to 400 ppm in 2014.

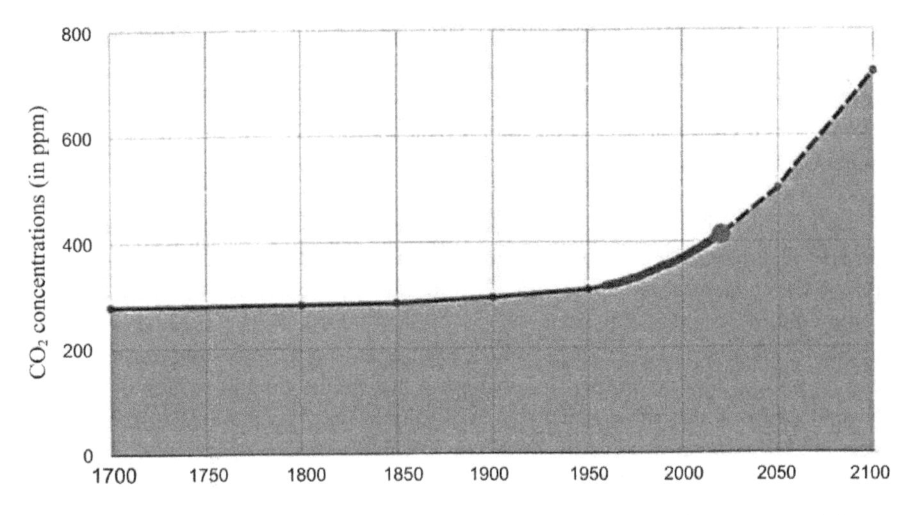

Figure 1.1. *Evolution of CO2 concentrations (in parts per million, or ppm) in the atmosphere since 1700, with a dotted line to predict levels up to 2100 according to the worst-case scenario provided by the IPCC scenario 8.5*[1] *(based on NOAA/ASRL/ SIO/IPCC)*

Figure 1.2 illustrates how global CO_2 emissions evolve (in gigatons, or Gt) between 1960 and 2020 (shown in red), distinguishing those produced due to fossil fuels and industry (shown in gray), from those produced due to occupation soils (shown in green). It can be concluded that the emissions growth is essentially due to the consumption of fossil fuels and industrial activity. We can also note the effect of the 2009 economic crisis, which reduced industrial activity overall. In 2020, we can see that CO_2 emissions drop to 39.94 Gt compared to the 43.06 Gt emitted in 2019,

1 Available at: https://www.notre-planete.info/indicateurs/CO2-dioxyde-carbone-concentration.php.

that is, a drop of more than 8%. This is mainly due to the slowdown in economic activity induced by the Covid-19 pandemic.

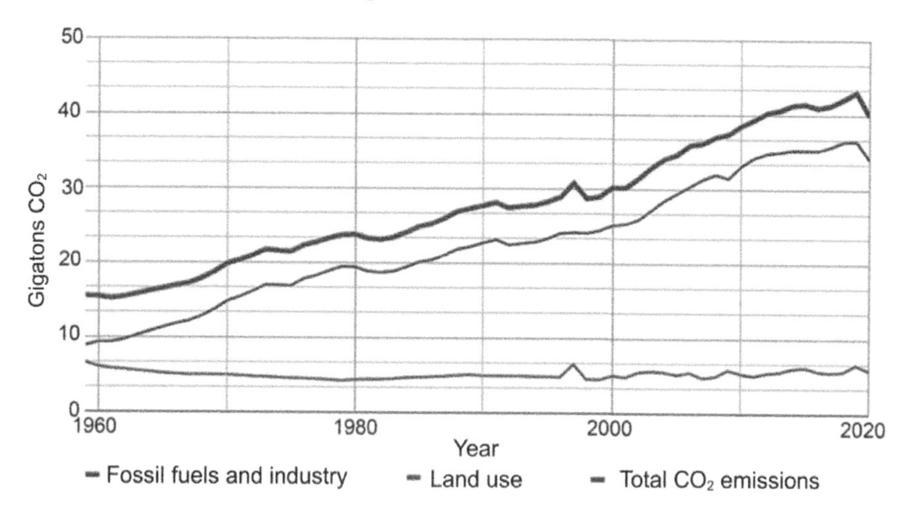

Figure 1.2. *Evolution of global CO_2 emissions (in gigatons, or Gt) between 1960 and 2020 (shown in red), those due to fossil fuels and industry (shown in grey) and those due to land use (shown in green)*[2]*. For a color version of this figure, see www.iste.co.uk/robyns/smartusers.zip*

1.2.3. *Renewable energies, or almost renewable energies*

Energy has always been present in the universe in two ways: nuclear energy contained in matter and kinetic energy created by the movements of stars and other particles moving in space.

The atoms that make up all matter possess enormous energy proportional to their mass. The sun and nuclear power plants harness this energy. Part of the heat in the center of the Earth results from nuclear reactions which allow for deep geothermal energy.

The sun gives off intense electromagnetic radiation which results from an enormous amount of nuclear fusion reactions. The Earth captures some of this radiation, allowing it to heat up. It enables photosynthesis, which makes plants grow, produces biomass and thus life. Fossil fuels come from transforming organic matter over millions of years. The sun also enables shallow geothermal energy, as

2 Available at: https://www.notre-planete.info/terre/climatologie_meteo/changement-climatique-GES.php.

well as the production of photovoltaic electricity and heat by means of heat exchangers, the evaporation of water which produces hydraulic energy, and it also heats up moving air masses, enabling wind power.

The gravitational pull of the Moon revolving around the Earth creates the tides, and the energy created from this process can be harvested by tidal power stations. Part of the wind comes from the Earth's rotation on itself [CAS 20].

In our times, renewable energies are those that can be continuously provided by nature. In this way, they come from solar radiation, the earth's core and the gravitational interactions of the Moon and the Sun with the oceans. A distinction can be made between different renewable energies – that is to say, wind, solar, hydro, geothermal and biomass origin [ROB 21].

Fossil fuels obviously do not fall into this category, because they are consumed in a much shorter time (less than 200 years) than what is required for them to be created (which is several million years).

Nuclear energy does not generate CO_2, with the exception of CO_2 emitted during the construction and deconstruction of power plants and when enriching uranium that is consumed in power plants. Uranium reserves are approximately 90 years long, based on "reasonably assured resources" which are added to "additional recoverable resources", at less than $130/kg, at a cost of between $80 and $260/kg of natural uranium and conventional fission which exploits the 235 isotope. This type of energy, which is therefore not a renewable source, will continue to be developed in a certain number of countries including France, subject to a satisfactory processing and management of waste, the development of a new generation of safer reactors, as well as, in the long term, the development of nuclear fusion. The latter concept will only be properly explored well beyond the year 2050 [ROB 21].

It is important to note that hydrogen is not a renewable energy source, because even though it is the most abundant atom in the universe, it does not exist in an H_2 form that can be directly used on Earth. It must be extracted from methane, which is fossil energy, or water by electrolysis, which consumes electrical energy. However, when used as energy, hydrogen does not emit CO_2 if it is produced from renewable or nuclear energy, other than that generated during the production of these technologies.

Figure 1.3 provides information about the estimated amount of greenhouse gases, essentially CO_2, per ton of oil equivalent (1 toe = 11.6 MWh) of final energy, that is, useful energy, for various energy sources.

Table 1.1 shows the quantities of CO_2 that are emitted into the atmosphere by different sources that are used to produce electricity. These emissions quantified in g/electric kWh were obtained via an analysis of the life cycle (LCA) of the sector which combines the construction of the installation, the extraction and transportation of fuel, the production of electrical energy, the treatment and storage of waste and the deconstruction of the installation.

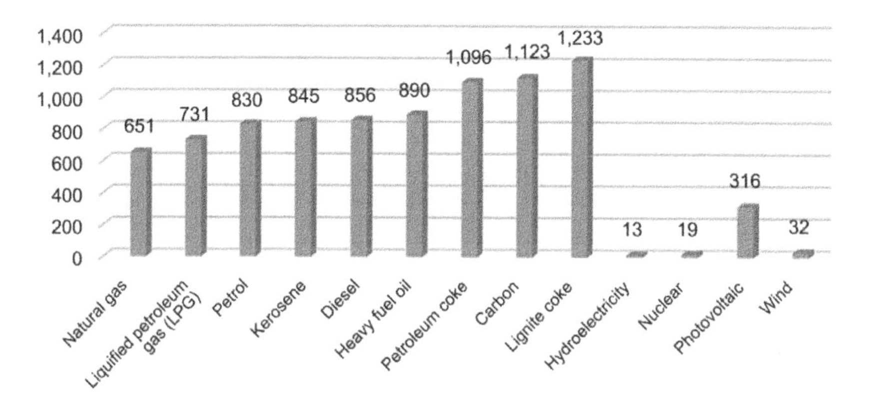

Figure 1.3. *Estimated greenhouse gas amount, mainly CO_2, per ton of oil equivalent (toe) of final energy for diverse energy sources [JAN 20a]*[3]

Transportation between the place where components are manufactured and where the power plants are installed has a significant impact in the case of wind power, and especially photovoltaics. High amounts can be obtained in the case of photovoltaics if battery storage is associated with the plant. This means one must take the LCA of the battery into account when evaluating the overall LCA of the photovoltaic energy source.

The value of wood energy emissions given in the table is extreme. If the forest from where the wood comes is managed sustainably, by replanting as much wood as is extracted, emissions induced by the use of wood energy are low, generated only by the cutting and transportation of wood.

Finally, Table 1.1 shows that an adequate and sustainable implementation of renewable energies allows for the production of low-carbon electricity. Apart from renewable energies, only nuclear energy has low carbon emissions. However, this type of energy also raises the question of radioactive waste and carries risk. In the future, the use of coal, which remains abundant on the planet, could still be

3 Available at: https://jancovici.com/changement-climatique/economie/quest-ce-que-lequation-de-kaya.

implemented through the recovery and storage of CO_2 emissions, a technology which is currently in development, but which is costly, and leaves the issue of secure storage of CO_2 open.

Electricity production source	CO_2 emission in g/kWh
Carbon	800 to 1,050 depending on the technologies
Petrol	985
Gas combined cycle	430
Nuclear	6
Hydraulic	4 to 70 depending on the type of unit
Wood biomass	200 to 1,500 without replanting (deforestation)
Wind	3 to 22 depending on the place of manufacture
Photovoltaic	5 to 150 depending on the place of manufacture and the type of cell

Table 1.1. *Amount of CO_2 emitted in the atmosphere by different sources which enable the production of electricity, in g/kWh, determined from an LCA*[4]

1.2.4. *Energy and economy*

Humanity has developed significantly because of energy, which is a development that has progressively accelerated for about 200 years with the exploitation of fossil fuels, which still correspond to 80% of energy currently consumed. This is an extremely short time scale compared to the millions of years that it took for *Homo sapiens* to appear and the 4.5 billion years that constitute the age of planet Earth. An abundance of energy has allowed mankind to work less for food, housing, treatment, etc., allowing for reflection, developing culture and creating new knowledge and new technologies. A significant correlation exists between higher energy consumption and longer life expectancy. A correlation also exists between energy consumption and the United Nations Human Development Index, which takes into account life expectancy, level of education and per capita income, and the Human Capital Index of the World Bank, which considers a child's chances of survival until age 18, the duration and achievements of their school education and their health [CAS 20].

In economics, the gross domestic product (GDP) is widely used to measure economic growth and therefore human activity (though this is not necessarily a

4 Available at: https://jancovici.com/changement-climatique/quel-monde-ideal/existe-t-il-des-energies-sans-co2 and https://bilans-ges.ademe.fr/documentation/UPLOAD_DOC_FR/index.htm?renouvelable.htm and EDF.

reflection of human happiness!). Economic activity requires a lot of energy to extract raw materials, transport them, transform them, provide services, etc. World GDP truly began to increase in the middle of the 19th century in connection with coal consumption, and the evolution of GDP has since been exponential. Figure 1.4 compares consumption of primary energy (energy consumed before transformation) with GDP on a global scale. It shows that the greater the GDP of a country, the higher the level of primary energy consumption, and vice versa. The relationship is more or less linear, as the line shows. There is therefore a very strong correlation between wealth creation (measured by GDP) and primary energy consumption.

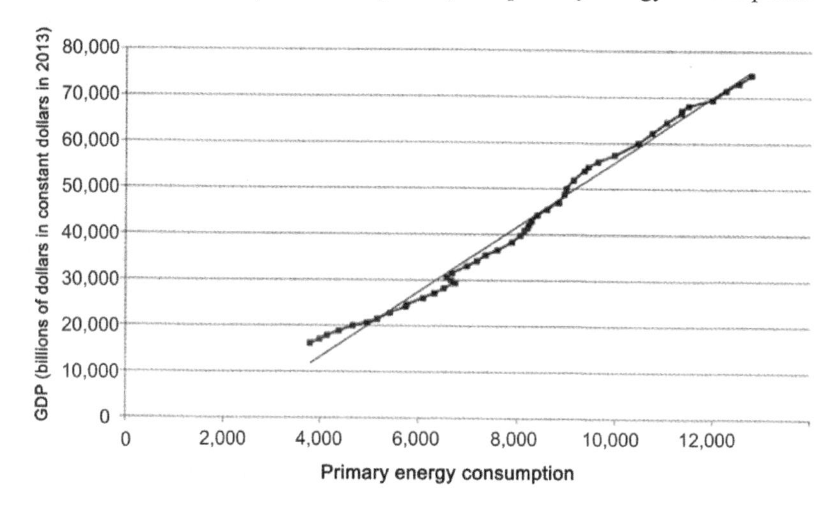

Figure 1.4. *Global GDP evolution in constant dollars (in 2013) in relation to primary energy consumption in Mtep [GIR 14]*[5]

In 1993, the Japanese economist Yoichi Kaya proposed an equation that highlights the links between CO_2 emissions, demography, GDP and energy [CAS 20, JAN 20a]:

$$CO_2 = Population \times \frac{GDP}{Population} \times \frac{Energy}{GDP} \times \frac{CO_2}{Energy}. \qquad [1.1]$$

The first term represents the world population, which came to 6.9 billion people in 2010, almost 8 billion people in 2020 and which could reach 9.8 billion in 2050 according to the United Nations (UN). The delta between 2010 and 2050, ΔP_{10-50}, is therefore 1.42.

5 Available at: https://climatetenergie.wordpress.com/2018/02/20/energie-croissance-economique-et-energie.

The second term represents wealth per person, in other words purchasing power. Of course, the majority of the population would like this term to increase! Between 2010 and 2018, it increased by 1.61%. Following this same pace, the increase, ΔGDP_{10-50}, will be 1.9% in 2050.

The third term illustrates the energy intensity of GDP, shown in Figure 1.5. Its evolution over time reflects the energy efficiency of our systems and processes, such as our habitats, electricity production, industries, transport.

Fortunately, if primary energy consumption grows at the same time as GDP does, it grows less rapidly. Between 2010 and 2018, the energy to GDP ratio decreased by 1.26% per year. By extrapolating this observation, we can obtain a delta, ΔE_{En_10-50} of 0.6 between 2010 and 2050.

Finally, the fourth term represents greenhouse gas emissions, mainly CO_2, that are emitted by the primary energies consumed. It will be at its weakest when exploiting renewable energies or nuclear energy. This ratio only decreased slightly (by 0.45%), between 2010 and 2018. This is because while the use of renewable energies has increased, the use of coal has also increased in certain regions of the world in order to meet growing demand faster than renewable energy sources. By extrapolating the variation between 2010 and 2018 up to 2050, we obtain for this ratio a delta, ΔE_{CO2_10-50} of 0.84.

By multiplying the estimates for the four terms on the right of the equation, we obtain the following result:

$$CO_2\ (2050) = \Delta P_{10-50} \times \Delta GDP_{10-50} \times \Delta E_{En\_10-50} \times \Delta E_{CO2\_10-50} = 1.42 \times 1.9 \times 0.6 \times 0.84 = 1.36.$$

This means that CO_2 emissions would increase by 36% by 2050. However, the IPCC recommends that CO_2 emissions decrease by 91% in 2050 compared to 2010 to limit global warming to 1.5°C in 2100 [INT 18].

Starting from the hypothesis that the first two terms cannot be reduced, since economic decline is not favorable among the societal majority, the last two terms of the equation should be drastically reduced as follows (the values of 0.183 are assumptions allowing the target of 0.09 to be reached for the evolution of CO_2 emissions):

$$CO_2\ (2050) = 1.42 \times 1.9 \times 0.183 \times 0.183 = 0.09.$$

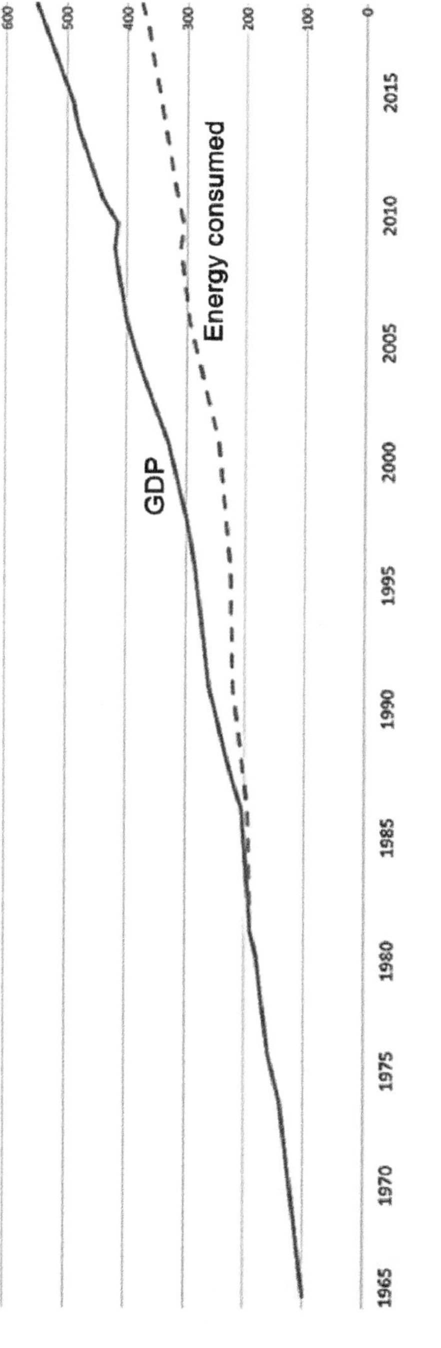

Figure 1.5. *Evolution over time of world GDP in constant dollars in 2010 and primary energy consumed in the world, with base 100 in 1965 [CAS 20]*

We should considerably increase the efficiency of our energy systems (gain even more than 40% efficiency) and almost exclusively use energy sources that do not generate greenhouse gases (reduce the use of carbon-generating energy sources by 65%). A decrease in the second term would mean consuming energy in a sober manner by modifying our lifestyles, even at the risk of limiting our comfort, without reducing the efficiency of our health systems. Even if an evolution toward more sober choices and behaviors is desirable, a drastic reduction of this term seems difficult to imagine, considering that nearly a billion human beings do not yet have access to electricity. However, limiting the main sources of energy that are currently exploited, as well decreasing the raw materials available on planet Earth, will one day inevitably lead to limited growth of the first two terms of the equation. These reflections do not account for the effects of a pandemic which would automatically decrease these first two terms (the Covid-19 pandemic in 2020 reduced GDP of France by around 8% and global GDP by around 5%)…

Many people are announcing serious economic and social crises induced by climate change, and yet solutions exist that will make it possible, if not to avoid these crises, to reduce their effects.

1.2.5. *Energy and meaning*

"I believe in an afterlife, simply because energy cannot die; it circulates, transforms and never stops". These were the words of Albert Einstein. Einstein showed that matter can be transformed into energy (atomic energy) and vice versa (as at the very beginning of the life of the universe) using the famous formula $E = mc^2$ (m being the mass and c the speed of light). Matter is in fact only a gigantic concentration of energy in an insignificant volume. There is also a link between energy and information. Information is contained in the genetic code, and this requires a lot of energy in order to create living forms. Information is needed to shape energy, followed by matter. Information is carried by energy, for example via radio waves. Information then becomes abundant and the energy required to process it becomes significant, as we will see later in this book.

Energy is therefore essential for human life. However, the combustion of fossil fuels not only contributes to a huge increase in greenhouse gases and therefore to global warming, but also generates pollution directly, which is a source of death. Coal-fired power plants cause nearly 23,000 premature deaths every year in Europe (Poland 5,380; Germany 4,350; Britain 2,860; France 390, etc.) [WWF 16]. Note that 670,000 people died in 2012 from burning coal in China. The cost of atmospheric mortality is estimated in China at 10% of GDP [GRA 16]. The number of premature deaths due to global pollution in France is estimated at 50,000 per year. This means that pollution kills more people today than tobacco [BAR 20].

The dangers of nuclear energy are better known to the general public, because they are more sensational and more publicized. The Chernobyl and Fukushima accidents are emblematic examples of this. But, according to the World Health Organization, fossil fuels account for several hundred Chernobyl deaths each year [CAS 20].

Figures 1.6–1.8 illustrate the number of illnesses and deaths induced by electricity production alone in Europe [MAR 07]. Figure 1.6 shows the number of fatalities per accident per 100 TWh of electricity produced. Figure 1.7 represents the number of deaths due to pollution per 100 TWh of electricity produced. In addition, Figure 1.8 highlights the number of serious illnesses. It should be noted that electrical energy consumed annually in Europe in 2018 was around 3,330 TWh. Annual electricity consumption per capita in Europe is high, at 5,448 kWh in 2016, 75% higher than the world average consumption [WIK 20a]. Among the energy sources taken into account in Figures 1.6–1.8, nuclear energy appears to be the least dangerous. However, renewable hydraulic, wind and photovoltaic energies have not been taken into account. According to Rabl and Spadaro [RAB 01], wind energy is the least lethal energy, followed by nuclear energy. Hydraulics do not generate direct pollution, but accidents caused by the rupture of large dams have already caused deaths, such as the rupture of the Banqiao dam in 1975 in China, which is said to have caused more than 100,000 deaths [CAS 20].

These illustrations of the mortal dangers presented by implementing and using energy sources raise the question of the value attributed to them, independent of their economic value. Indeed, we seem to attach less importance to risks that seem more diffuse, but more intense in terms of mortality, than to risks that are more impressive, but lower in intensity. Risks that are diffuse and random in nature seem more acceptable. This finding will no doubt lead some countries to pursue the development of nuclear energy, even in a moderate sense, as well as to a drastic reduction of this type of source in other countries. We also cannot lose sight of the treatment and storage of highly reactive waste over long periods of time, which can raise ethical questions, since it is a question of leaving waste management to future generations. In addition to this, global warming tends to increase the temperature of cold sources (rivers, seas, air) which are necessary for nuclear power plants to function properly. This means risking a loss of productivity in order to ensure their safety.

The value of energy also comes from its accessibility to all, insofar as we can see how it enables a standard of living with decent levels of health and education. As a reminder, nearly a billion people do not yet have access to electricity. In addition, 14% of the French population finds themselves in fuel poverty, that is to say, they have difficulty paying their electricity and heating bills, which can reduce their access to these energy sources. We must therefore ensure that the essential energy transition does not overlook these populations, rather that it allows everyone a minimum level of access to different sources of energy, by implementing modes of energy solidarity and, where appropriate, local solutions and regulation of the forms of competition that would be unfavorable to this transition. According to Nicolas Hulot, the economic and ecological transition must allow countries and populations to strike an economic balance, to erase injustices and to forge links that will produce peace and solidarity [GRA 16].

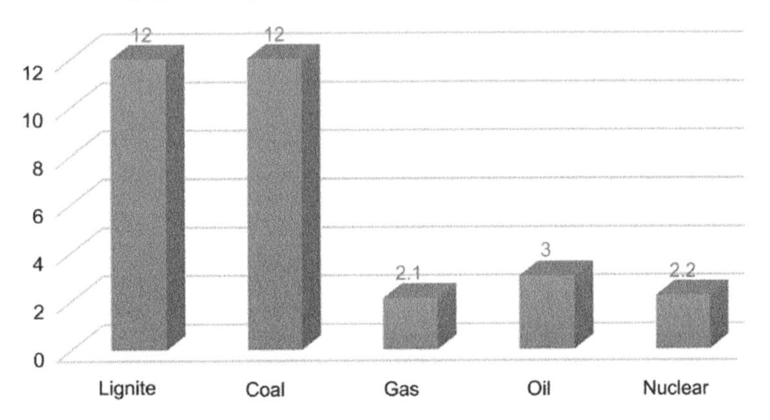

Figure 1.6. *Number of accidental deaths for 100 TWh of electricity produced [MAR 07]*

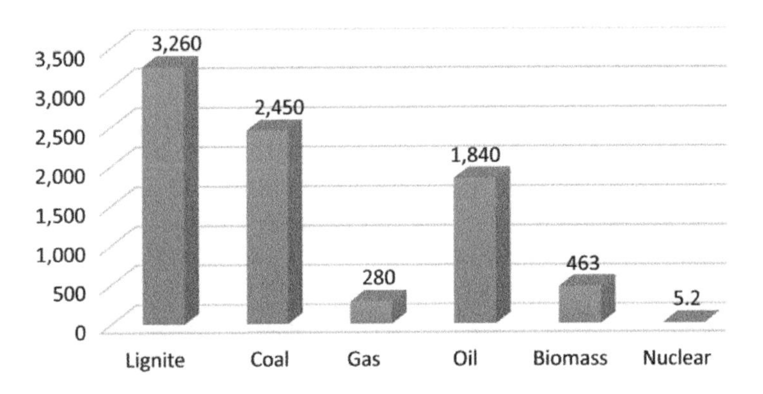

Figure 1.7. *Number of deaths caused by pollution for 100 TWh of electricity produced [MAR 07]*

Beyond energy poverty, the energy and societal transition must be inclusive by involving all populations: poor, rich and well off. However, beyond the financial question, we must allow everyone to access knowledge about climate and societal challenges, to access and appropriate technologies that can help the transition, to be able to take the necessary steps in order to understand the world differently and to change one's relationship to the world in order to preserve it. This is for the good of everyone, but especially of future generations.

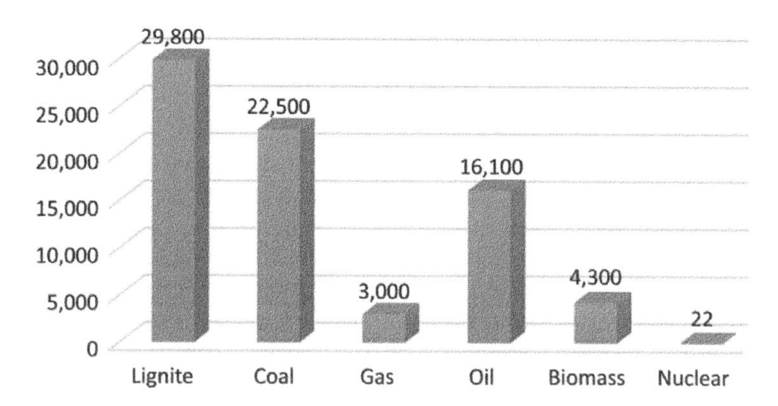

Figure 1.8. *Number of serious illnesses due to pollution for 100 TWh of electricity produced [MAR 07]*

1.3. Opinions surrounding climate change

1.3.1. *Denial and inaction*

The climate emergency appears increasingly in the news. The climate changes already felt by everyone will have impacts in many senses – environmental, societal, health, economic, etc. And yet, a level of denial continues to persist at times among leaders of major world powers, which is all the more worrying. This is an attitude that can lead to withdrawal when faced with the scale of the phenomenon. This withdrawal into oneself cannot however be sustainable as the economy becomes increasingly globalized, as no country will be spared the effects of global warming.

The economist Nicholas Stern, former chief economist at the World Bank, has pondered the cost of climate change: "If left unaddressed, the overall costs and risks of climate change will be equivalent to the loss of at least minus 5% of global GDP every year, now and forever. If a wider range of risks and consequences are taken into account, damage estimates could amount to 20% of GDP or more. On the other hand, the costs related to action, namely the reduction of greenhouse gas emissions to avoid the worst consequences of climate change, can be limited to 1% of global GDP each year" [GRA 16].

In 2013, the World Bank alerted the world to the cost of natural disasters, which are set to increase with climate change. It estimated that natural disasters have killed 2.5 million people and cost 3,800 billion dollars between 1980 and 2012, mainly in emerging countries, which have seen their GDP weakened by climate change [GRA 16].

1.3.2. *Faith in technosciences*

A second view is based on total faith in technosciences. Science and technology will save us, and tomorrow we will find new sources of energy and new technologies.

One could postulate that the scarcity of fossil fuels will force us to do just this. In this way, we can count on a shortage to limit climate drift. However, in order to limit the increase in temperature to 2°C in 2100, it is essential that we do not emit more than a thousand gigatons of CO_2 by the end of the century. Of this thousand gigatons, fossil fuels are affected by 300 gigatons of carbon. But knowing that the latter emit a little more than 30 gigatons per year, at this rate, the carbon budget will be largely exceeded before 2050, especially since the known reserves of fossil fuels allow for it [GRA 16].

Technology is largely at the origin of the climate problem, and it will certainly be part of the solution, but not alone. According to Pierre Giorgini: "What we are experiencing is more than a crisis, we are experiencing a dazzling transition from an old world to a new world. This transition is economic, financial, social, environmental and geopolitical. It is global and is part of a transformation of the world whose nature no one is able to accurately predict. So it's a transition to something else. The strength of this transition is probably unprecedented in terms of both its acceleration and its magnitude. This apparent magnitude comes in particular from the combination of a new techno-scientific revolution, with the shift to a new paradigm relating to modes of cooperation between man and machines, combined in turn with the transition to the creative economy. We are in fact talking about a systemic upheaval" [GIO 14, GIO 16b].

Technology will be part of the solution, but how should one assess this part so that these technologies are effectively aimed at reducing the climate impact of human activities? Pierre Giorgini and Thierry Magnin [GIO 21a] propose an ethical approach to guide research and development actions by:

– promoting and participating in technological development when it aims to repair ecosystems, address ecological damage, the fight against pollution and global warming;

– favoring technologies when they repair the social fabric, contribute to reducing the digital divide when they aim to connect the human race around issues of safeguarding and the promotion of mankind in the protection of the freedom of conscience.

In this context, we must recognize the preferential option for vulnerable populations, and natural and social ecosystems that are in danger, as a universal value of law and conscience. Wherever we are engaged locally, for any development or technological innovation, regionally, nationally or internationally, in business or any civic activity, we must support, act and claim the option that favors the poor and the most vulnerable people, as a source of radical innovation that benefits all of humanity.

1.3.3. *Saving through economy*

Dominant thinking is liberal in nature: the market will be efficient, and invisible forces will come about when the time is right to integrate climate issues with financial parameters. Across a large part of the economic and financial world, investors would be perfectly informed. If stock market prices or insurance premiums do not reflect the risks linked to climate change, it is because these do not really exist. When risks materialize, investors will take them into account and financial flows will be oriented in such a way that we are able to deal with the new climate situation. There is no need to regulate financial authorities to direct flows toward the financing of the transition, as this will be done naturally in due time. However, the financial crisis of 2008 proved that perfect information held by the actors of the financial world is a myth [GRA 16]. The same is true for the pandemic crisis of 2020–2021…

Moreover, the costs of renewable energies decrease over time, and it can be assumed that they will inevitably become cheaper than fossil energies. The same will no doubt be true for various technologies that emit low carbon, or even allow carbon to be captured before it is emitted into the atmosphere. One could then believe that future generations will inherit solutions that are more economical than those that exist today, and which will therefore be able to be implemented more easily. Placing the responsibility on those who come after us is however audacious and ethically questionable... .

1.3.4. *The reason for sustainable development*

In 1986, the concept of sustainable development was defined as follows: "meeting the needs of the present without jeopardizing the ability of future generations to meet their own needs".

This concept involves exploiting renewable energy resources, which are the only guarantors of a sustainable environment, while minimizing the environmental impacts associated with their conversion and the manufacture of their converters. Fossil fuels emerge as a finite and economically limited resource, inducing emissions that affect the environment and contribute to climate change. A sustainable energy system must integrate renewable energy sources and/or conversion chains using low-emission renewable fuels, which are accessible at acceptable costs. Despite the fact that the creation of new energy infrastructures takes several decades, a growing number of large companies are involved in the development and commercialization of these new technologies.

Sustainable development requires managing a balance between economic development, social equity and environmental protection in all regions of the planet. This concept cannot therefore come to pass without the real political will of a growing number of countries [ROB 21].

Pierre Giorgini expresses this concept in the form of an ethical rule, going further by emphasizing the need to restore original ecosystems: "You will not entrust your children with the task of solving problems that you have created voluntarily, which are vital for your offspring and for which you are not sure of the current or future existence of a realistic solution. On the other hand, the advances generated by scientific discovery and/or technical development will aim to strengthen the common good and will promote the restoration of the original ecosystems if they provide balance and harmony, wherever possible" [GIO 18].

The United Nations (UN) has set 17 Sustainable Development Goals to save the world, shown in Figure 1.9. These Sustainable Development Goals provide the roadmap for achieving a better and more sustainable future for all. They respond to the global challenges we face, including those related to poverty, inequality, climate, environmental degradation, prosperity, peace and justice. The goals are interconnected and, in order to leave no one behind, it is important to achieve each of them, and each of their targets, by 2030. These 17 goals are broken down into 169 sub-goals for which the contribution of each actor (company, community, association, university, etc., but also the individual) can be evaluated through indicators.

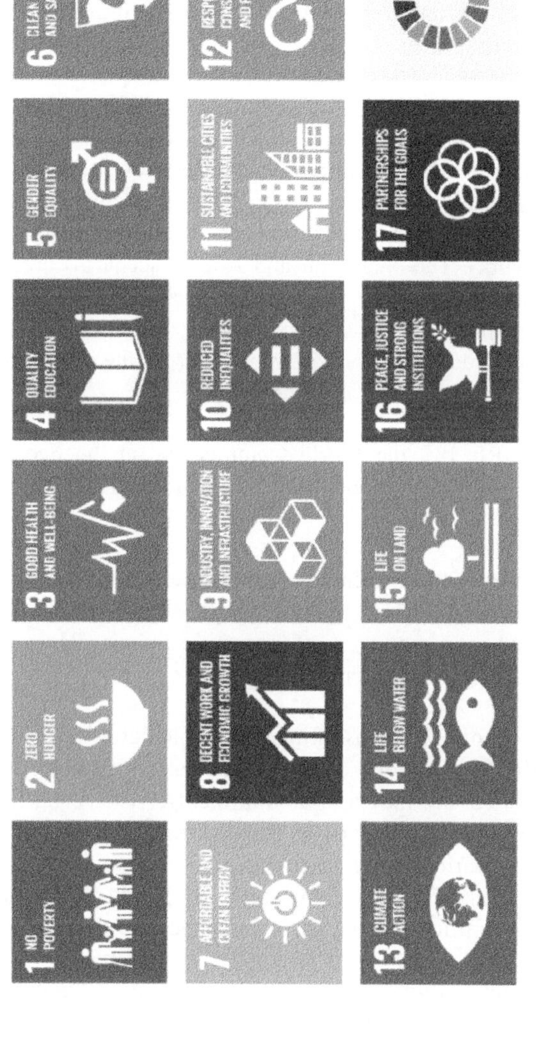

Figure 1.9. *The 17 sustainable development objectives from the UN*

1.4. Scenarios and possible solutions

1.4.1. *Scenarios, so many stories of a desirable future?*

In order to limit the increase in temperature in 2100 to 2°C, we should aim for carbon neutrality by 2050. Carbon neutrality does not only mean that we no longer emit carbon at all, but that the carbon being emitted is either offset by the development of carbon sinks (forests), or it is captured and stored so that it is not released into the atmosphere.

With the aim of achieving carbon neutrality by 2050, various organizations are proposing scenarios based on the evolution of technologies, the economy, governance and territories, but also of our lifestyles. In France, in 2021, three organizations proposed scenarios: NégaWatt, Ademe (French agency for environment and energy management) and RTE (manager of the electricity transmission network). The NégaWatt and Ademe scenarios present proposals that can be widely shared on a global scale. The scenarios proposed by RTE are strongly inspired by the current energy mix in France, which relates to a study of an electricity grid powered by 100% renewable energy, the conclusions of which can be extended to many countries around the world.

Greenpeace [GRE 13], the World Commission on the Economy and Climate [NEC 15], and the American economist Jeremy Rifkin [RIF 12] offer scenarios with an international vision.

The main aspect of these scenarios is sobriety, which is strongly linked to our lifestyles, an increase in the energy efficiency of our systems (housing, transport, industries, etc.) and an increased use of renewable energies. However, the weights of these axes differ according to scenario concerned.

These scenarios form the basis of many possible narratives for the next 30 years. It is desirable that at least one of them materializes in order to achieve carbon neutrality in 2050, but are they desirable in themselves? A big unknown relates to the acceptability of these scenarios by populations, in particular those calling for sobriety. These scenarios generally implicate a social change, an evolution of behavior, even a break in everyone's vision of the world and in their relationship to it. It is therefore a significant challenge to make these scenarios desirable…

In the rest of this section, the trending scenarios are presented, though they are by no means exhaustive. All these scenarios can be explored further via the websites of the organizations that offer them. These scenarios and others are beginning to be questioned on certain particularly important aspects, such as the role of digital technology, the importance of returning to the local, the involvement of all actors and the development of demonstrators.

1.4.2. *Renewable energies and sobriety*

The NégaWatt scenario aims for carbon neutrality in France by 2050 and the reduction of greenhouse gas emissions by 55% on a European scale by 2030 compared to 2019. Figure 1.10 illustrates the evolution of carbon emissions in France until 2050 by considering a trend evolution (shown in red) which assumes that few intentional reduction actions are implemented, compared to the situation in 2020, and the evolution that aims to achieve carbon neutrality in 2050 (shown in blue).

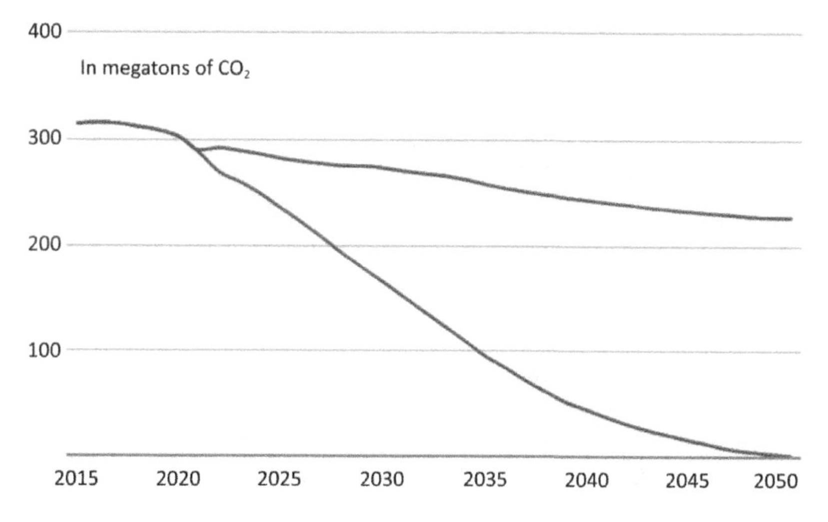

Figure 1.10. *Evolution of carbon emissions in France until 2050 considering a trending evolution (shown in red) and the objective carbon neutrality (shown in blue) [NÉG 21]. For a color version of this figure, see www.iste.co.uk/robyns/smartusers.zip*

The objective of reducing carbon emissions presupposes a sobriety of demand, by dividing into two the final energy consumption in France by 2050 (production of electricity, mobility, heating of buildings, industries). But according to NégaWatt: "The transformation of our energy system which is necessary cannot be achieved by simply replacing fossil fuels with carbon-free energy sources. Limiting environmental and social impacts and reducing the pressure on raw materials requires a profound transformation of the way we consume and produce energy and material goods. These perspectives call for strong societal transitions, both individually and collectively" [NÉG 21].

Sobriety is therefore one of the three major areas to be developed. According to NégaWatt, it will be necessary to prioritize the most useful uses, to restrict the most complicated and to eliminate the most harmful. Sobriety has several dimensions, partly conceptualized by the NégaWatt association [VIL 18, WIK 20b], in the spirit of democratization and sharing:

– dimensional sobriety: by favoring equipment that is adapted to the need, when choosing a purchase or an investment (e.g. opting for a smaller surface area for your living accommodation, using a vehicle adapted to the load and number of passengers);

– cooperative sobriety: by pooling uses, whether it concerns spaces, goods, etc. (e.g. carpooling, car sharing, co-location, loan of equipment between neighbors);

– sobriety of use: by managing the use of appliances and goods reasonably (e.g. eco-driving, use precautions to limit breakage and premature wear and tear of goods, regulation of heating);

– organizational sobriety: by structuring activities differently in space and time (e.g. promotion of telework, regional planning, provision of public transport);

– material sobriety: by reducing the consumption of material goods and products (e.g. reducing the rate of equipment, limitation of packaging). These same goods indirectly require energy to be designed, assembled, transported, etc. In this sense, we are thinking about indirect energy, or embodied energy.

The second axis is to increase the energy efficiency of our systems, that is, to reduce the quantity of energy necessary to satisfy the same need. This translates into better insulation of buildings, which accounted for 45% of overall energy in France in 2016 [ROB 19], through improved efficiency of electrical appliances and transport systems, etc.

Figure 1.11 illustrates the evolution of energy uses in France up to 2050 in the NégaWatt scenario. The gains obtained in energy efficiency when it comes to consumption and production constitute a significant part of the reduction in primary energy used up to 2050, that is, 38.2%, and energy sobriety contributes to 13.8% of the reduction in this energy.

The third axis assumes that energy production in all sectors (heat, electricity, transport) would be based 100% on renewable energies.

Figure 1.12 shows the evolution of CO_2 emissions by activity sector in France by 2050 according to the NégaWatt scenario. The most drastic reductions concern the energy sectors, manufacturing industries, transport and residential, tertiary, institutional and commercial buildings.

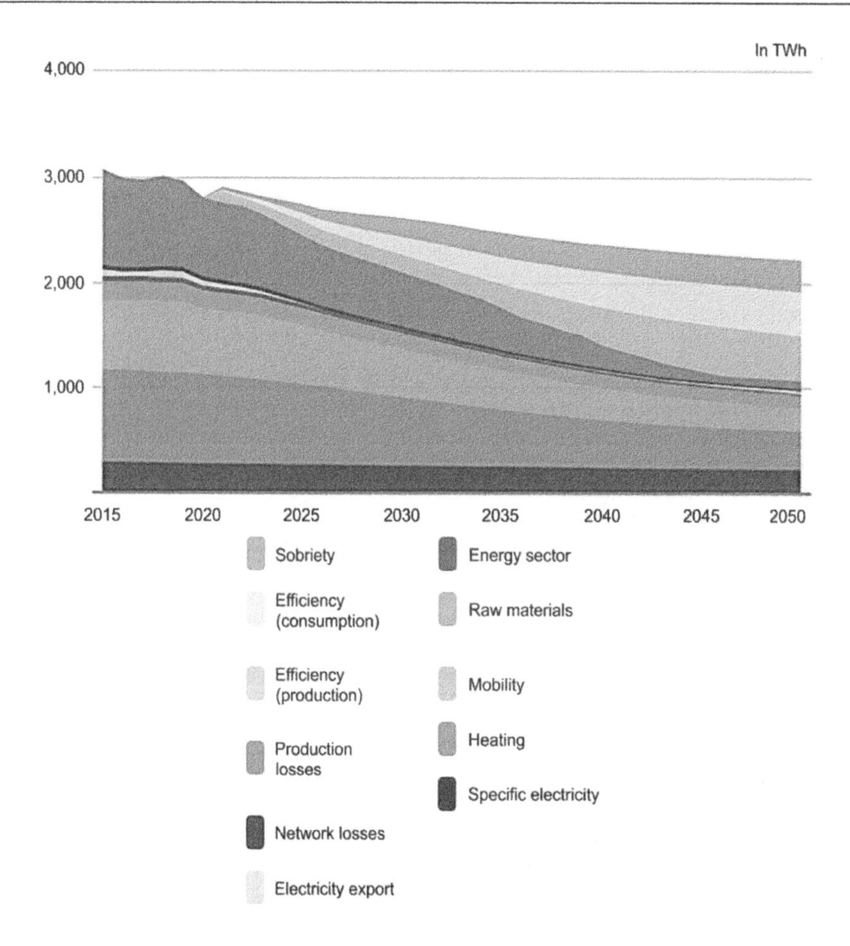

Figure 1.11. *Evolution of energy uses up to 2050 in France in the NégaWatt scenario [NÉG 21]. For a color version of this figure, see www.iste.co.uk/robyns/smartusers.zip*

The first scenario proposed by Ademe (S1), entitled "frugal generation", relies heavily on sobriety and energy efficiency [ADE 21]. This scenario is based on a search for meaning which leads to a sense frugality that has been agreed to (but also partly constrained by a legislative body), a preference for local products and nature which is protected. It advocates for society to divide the consumption of meat by three, to sharply reduce the mobility rate by making people take half the number of their journeys by foot or bike, and to renovate 80% of housing stock in low-energy buildings. Alongside this, it favors governance that prioritizes local decision-making, the creation of new economic indicators (based on income disparities, quality of life, etc.), the development of production processes which meet needs as much as possible and the promotion of a circular economy (a high rate of material recycling).

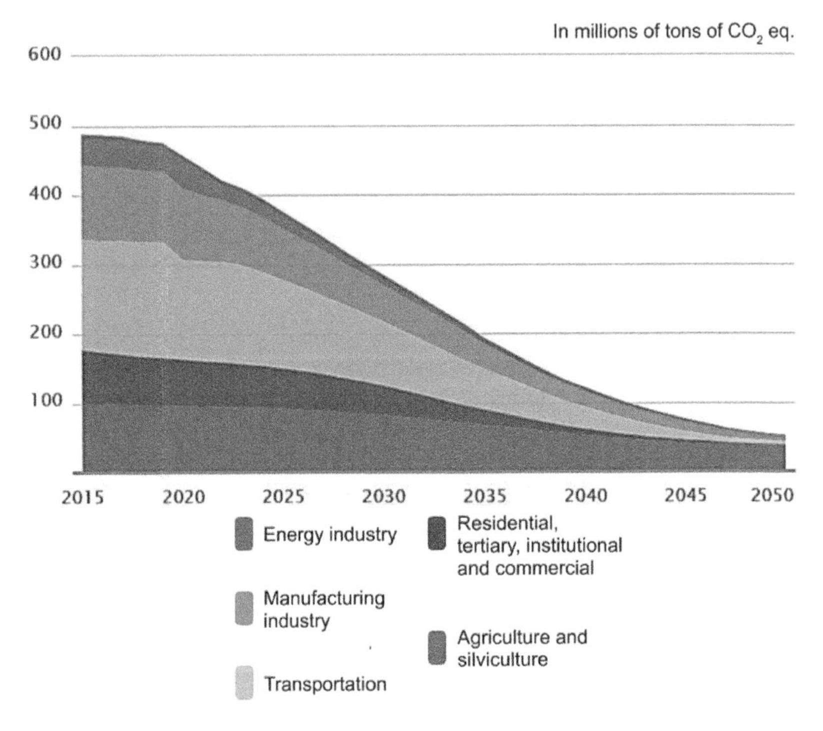

Figure 1.12. *Evolution of energy uses up to 2050 in France in the NégaWatt scenario [NÉG 21]. For a color version of this figure, see www.iste.co.uk/robyns/smartusers.zip*

Greenpeace also proposes scenarios based on sobriety and 100% renewable energy [GRE 13].

To keep up with the rate of growth envisaged for renewable energies and energy efficiency, it would be satisfactory to invest in the large sums required, provided there is support from the political bodies.

The question of sobriety is more delicate, because in addition to changing our lifestyles, it could be associated with economic decline, which would not be welcome globally. It is important to note that UN Sustainable Development Goal 8 (Figure 1.9) targets economic growth, and not decline. Another aspect to consider is reviewing the founding criteria on which our economy is based (such as GDP) in order to introduce new criteria such as quality of life.

Surveys carried out by Ademe show that in 2020, 58% of the French population were convinced that it would be necessary to change their way of life. According to 52% of French people, we must get out of the "myth of infinite growth" and completely review our economic model [ADE 20].

How might we make this evolution desirable? How might we find a new social consensus and modify collective imaginations to move from narrative to action? All of this without producing an increase in inequalities between the richest and the less well-off? A decline in fossil fuels and the necessary protection of biodiversity will force people to consume less and differently. In any case, lifestyle changes will be necessary, and "acceptable sobriety" remains to be seen [FRA 15].

1.4.3. *From 100% renewable energies to a mix of solutions*

In 2021, the French electricity transmission system operator published its six energy progression scenarios for France up to the year 2050 [RTE 21]. The scenarios center around electricity, but an essential aspect of all the scenarios is the story of possible futures. Since global energy consumption must decrease, electricity consumption will increase as fossil fuels are replaced in several sectors: transport, including electric vehicles, industry, hydrogen production from renewable sources, etc. This increase would sit at 35% in the year 2050 compared to 2020.

However, the report insists on a reduction of global energy consumption because of both an increase in the energy efficiency of energy systems, and to the concept of energy sobriety, which for some constitutes the first response to the environmental crisis, while others reject the principle in the name of individual freedom and to maintain a subjective form of "comfort" [RTE 21]. This decline is estimated to be around 40% in the year 2050 compared to 2020, but new European objectives aim to accelerate this decline to 55% by 2030.

A scenario based on 100% renewable solutions for the sole production of electricity has been proposed by Ademe [ADE 18] as well as by RTE [RTE 21]. Given the fluctuating nature of several renewable energy sources – wind, photovoltaic, small hydraulics and marine energies [ROB 21] – it will be necessary to considerably develop how electrical energy is stored [ROB 15, ROB 17, ROB 19] and enhance the flexibility of loads, that is to say, to modulate electricity consumption over time, in particular by adapting to the variability of renewable sources, but also by investing in electricity networks adaptations. RTE imagines different variants of this scenario depending on the amount of development of onshore and offshore wind power and photovoltaics. One scenario predicts that one in two houses in France would be equipped with photovoltaic panels for partial self-consumption [ROB 19].

However, RTE also foresees scenarios whereby part of the electricity in France would still be generated by nuclear power sources, by extending existing power plants and building new ones. According to RTE, these scenarios would be less costly in terms of investment than a 100% renewable scenario. However, the uncertainties around the extension of certain power stations and the delays inevitable when constructing new power stations may lead to a situation in which the desired objectives are reached by 2060 and not by 2050. In this case, in order to achieve the objectives put forward for 2050 despite everything, the economic implications of the different scenarios would be similar.

In some cases, scenarios that are based on mixed energy solutions with renewable and nuclear energy sources offer the chance for some CO_2 to be captured which is emitted by the fossil fuels that are still in use, even though in steep decline, by certain industrial processes. These scenarios are, for example, proposed by Ademe [ADE 21] and the IPCC [INT 18].

Although still expensive, CO_2 capture and storage (CCS) will have to form part of the combination of solutions. The electricity sector is the largest emitter of CO_2 in the world, but it can get rid of CCS by altering production methods, in favor of renewable and nuclear energies, as is the case in France. This is not the case for countries that continue to exploit shale gas (like the United States) or coal (like China). But for industries such as steel or cement, the use of carbon-free energies only makes it possible to cut around half of the emissions generated by the industrial process; in these cases, CCS will often be the optimal solution. Processes to capture CO_2 are complex but well known and mastered by several industrial sectors. CO_2 is stored in saline aquifers situated around 1,000 m underground or under the surface of the sea. This is a technique mastered by the oil sector, in particular the chemicals industry. Such storage facilities already exist in Norway. Their difficulty lies in the large-scale deployment of this capture and storage technique, which is costly.

Alain Grandjean [GRA 16] advocates for the scenario published by the World Commission on the Economy and the Climate, also called NCE for New Climate Economy [NEC 15]. This scenario suggests favoring compact cities. The comparison of the cities of Atlanta in the United States and Barcelona in Spain illustrates this proposition. These two cities have comparable standards of living and populations with approximately 5.3 million inhabitants. Atlanta covers 4,300 km^2 while Barcelona occupies 162 km^2. In Atlanta, public transport and individual transport emit 7.2 tons of CO_2 per inhabitant per year, but Barcelona emits 10 times less. In Ademe's S2 scenario, entitled "territorial cooperation", we see the idea of favoring medium-sized towns over large agglomerations. These cities are becoming denser in a controlled manner and are becoming "quarter-hour cities" where almost everything is nearby, allowing for the development of soft mobility and public transport [ADE 21].

The NCE report also highlights the need for reforestation. About a quarter of the world's agricultural land is degraded, and climate change threatens water supplies. Net deforestation (meaning the balance between deforestation and reforestation) is responsible for approximately 11% of greenhouse gases. Reforestation is necessary to create carbon sinks, which are essential to gain time in the fight against global warming. Finally, as in all the other scenarios, it is proposed that all energy carriers are decarbonized. According to this report, decarbonizing the economy does not cost more than maintaining the status quo.

1.4.4. *The Third Industrial Revolution*

The first industrial revolution was that of coal in the 19th century, the second that of oil in the 20th century and the third that of digital technologies in the 21st century. The Third Industrial Revolution has been characterized by economist Jeremy Rifkin [RIF 12]: it is based on renewable energies, energy-producing buildings, energy storage in buildings, energy exchanges via a smart grid and electric vehicles. He is optimistic because he thinks that because of these solutions, society will be able to continue to consume energy without resorting to sobriety, possibly experiencing a drop in energy consumption for equivalent service where possible.

Rifkin stresses that the Internet of energy, materialized by mainly smart grids, which are associated with the Internet of Things (objects communicating with each other via sensors), will play a strategic role in what is to come. He speaks of an Internet of transport that can optimize the loading of trucks to avoid empty transport. He then speaks of Internet convergence. The Hauts-de-France region was largely inspired by his story. The region adapted it to their specific context so that they could launch their own industrial revolution which was aimed at transforming their economy while reducing their carbon footprint. This approach will be presented in Chapter 2 alongside some initial feedback.

Pierre Giorgini goes further by bringing the highly connected or interconnected human into the picture: "This internet convergence applies to all levels of our human activity. This dazzling transition therefore brutally modifies the place of mankind in organized systems, calling on them to bear more universal conscience in each of their actions within interconnected communities. Mankind is no longer just a source of data and information servers, they are both a source and a recipient of it. The very notion of an 'organized society' is misplaced. The exercise of subjectivity and the imagination may be disturbing for some and may also raise the question of the future of the subject" [GIO 14, GIO 16b].

1.4.5. *Smart due to digital technology*

Like the First Industrial Revolution profoundly changed society in the 19th century, the digital transformation we witnessed at the beginning of the 21st century is still in its infancy, but we already know that it will revolutionize our world. In fact, technological progress makes it possible to envisage a world where almost all the objects that surround us will interact not only with us humans, but also with each other. It is the Internet of Everything (IoE or Internet at all): gigantic computer networks where several billion human beings and tens of billions of real or virtual objects exchange massive amounts of data.

Big Data will allow artificial intelligence (AI) algorithms to assist users and actors within buildings, energy networks and cities in order to make them more economical and more sustainable, and to make them smart buildings, smart grids and smart cities.

But we will have to be vigilant because digital technology and AI are energy intensive. AI requires machines with rare materials for batteries, electrical power for operations and lots of energy to cool and manage hardware. However, solutions are emerging which allow one to recover the heat released by data centers and use it for other purposes. It will also be a question of thinking about the genuinely useful pieces of data, how frequently they are acquired and the operations that need to be carried out wisely.

In 2018, digital technology was the source of 3.7% of global greenhouse gas emissions, and 4.2% of global primary energy consumption: 44% of this footprint occurs due to the manufacture of terminals, computer centers and networks, and 56% to their use [SÉN 20].

Currently, the digital sector emits as much carbon as the air transportation sector. In 2019, the digital sector represented 2% of France's carbon emissions. Growth of up to 6.7% in 2040 is predicted if nothing is done to reduce the carbon impact of the digital sector [SÉN 20].

It should be noted that "the Internet of energy" holds special status within the IoE because of the challenge it represents in terms of the sustainability of the planet (regarding the management of random renewable energies, the control of consumption profiles, the increase in the energy efficiency of buildings, transport systems, etc.). This is a strategic issue that impacts the effective functioning of the economy and of security systems. Additionally, from a legal point of view, we must consider the binding regulations surrounding the right to produce one's energy, to use public energy networks, to meter energy and to share it. Indeed, new approaches are beginning to develop in order to make it possible to produce and consume energy locally, called individual and collective self-consumption in France, and more generally local renewable energy communities.

Ademe's S3 scenario entitled "green technologies" is based on optimizing performance in all sectors without radically changing behavior, because of digital technology which helps to create a more optimized and connected society. This scenario predicts that by 2050, data centers will consume 10 times more energy than in 2020. This energy will however be carbon free, and the CO_2 still emitted would be captured and stored [ADE 21].

Ademe's fourth and final scenario (S4) entitled "remedial bet" relies even more on digital technology and technology to maintain our lifestyles based on consuming goods and energy. It depends on society's ability to manage and repair social and ecological systems which are damaged by global warming. This approach is based, among other things, on the capture and storage of the CO_2 that is generated, but which is also already present in the atmosphere, because of technologies that are not yet technically and economically mature on a large scale. It also predicts an increase in the importance of digital technology, the omnipresence of the Internet of Things and AI, and data centers which will consume 15 times more energy in 2050 than in 2020 [ADE 21]. This is a trending scenario, similar to the current trend we can see in our society, which entails the significant risk of not achieving the necessary carbon neutrality by 2050. But is this a desirable narrative?

1.4.6. *From global to local*

Globalization has made many companies relocate, meaning that many countries have become dependent on only a select few states. This dependence had already emerged in the 1970s when it came to the supply of oil, a fossil fuel whose extraction is highly concentrated in the Middle East, a region that is relatively politically unstable. This is why some countries, including France at the forefront, have developed nuclear energy in order to reduce dependence. However, in recent times, a new dependence has emerged vis-à-vis countries in the Far East, including China and India, which begins to pose questions, even serious problems, for certain rare materials that are necessary for the manufacture of magnets which are used to construct electric generators for wind turbines. Examples of this are the manufacture of batteries, or of photovoltaic panels that make long journeys which generate carbon, or even for the supply of sanitary products, which is unacceptable in times of a global pandemic...

A movement is therefore emerging to bring essential productions from a global to a national scale. However, a trend is also emerging which favors even shorter circuits at the regional, urban or even district level. The development of local renewable energy communities and collective self-consumption goes in this direction. The same is true for short circuits for agricultural products, by eating locally, or even by sourcing from new urban farms that can produce food in any season. It is also a question of promoting a circular economy at the scale of a territory both in terms of recycling goods and reducing the distance of journeys, thereby reducing the carbon footprint.

In terms of everyday life and the relationship of the individual to economic and democratic activity, Pierre Giorgini believes, following the thinking of Jeremy Rifkin, "that technosciences that are emerging today will make a large decentralization possible, along with relocation on a domestic scale of energy resources, and the creation on a local scale of an energy mix, through inter-object cooperation within cooperative mesh networks known as smart grids".

In such a model, scarcity will no longer be at the center of the energy economy because energy will have become predominantly renewable or even renewed for solar, wind and bioenergy [GIO 16a]. He speaks of "glocality" to underline these links, or even the mesh that is to be recreated between the local and the global.

1.4.7. *All actors*

Smart grids will truly deploy their potential for innovation when promoting interactions between the various players in the electrical system (producer, consumer, storage company, network manager), themselves becoming "electrically" smarter players. These actors have very different consumption and/or production profiles, and extremely variable economic and societal objectives and/or constraints. New sets of actors can therefore appear, making it possible to target new economic models, but also to respond to energy and climate issues by promoting the development of renewable energies. One challenge tied to this evolution is that all the actors become winners, without leaving the actors who are experiencing fuel poverty to one side [ROB 19].

According to Erik Orsenna: "The time has come for consumers to take revenge on producers. The most striking example is that of energy. The old mode of production was in the hands of monopolies. No other source was available. The market was opened wide to many new players" [ORS 18].

Buildings are becoming smarter, as well as producing energy because of the deployment of numerous sensors and automations aimed at increasing energy efficiency and user comfort. This notion of comfort is very subjective and varies according to the users. Users are invited to become actors in their building, both to adjust their comfort levels, but also to contribute to the objective of energy efficiency. This is because technology cannot be used in an optimal way by everyone in the world, and also because it consumes energy itself. Ideally, smart buildings should be inhabited by smart users...

Thus, to succeed in the energy transition, everyone should become an actor working in their own interest, but also according to a logic of cooperation, thus making it possible to achieve objectives and overall gains which are greater than the sum of individual gains, acknowledging that the levels of rationality among individuals is very variable [ROB 19]. Pierre Giorgini speaks of cooperative mesh networks which allow one to generate a new form of creativity and new potentialities [GIO 14].

But, does everyone want to become an actor, consumer-actor (or *prosumer*)? In Giorgini and Vaillant [GIO 16b], Malik Bozzo-Rey, philosopher and ethicist, ponders the meaning of this term: "What does it mean for a consumer to behave like an actor? Are we not creating a huge illusion so as to ensure, precisely, that the consumer continues to perform only one action which is always the same? In other words, they do not stop consuming, which is after all the great anxiety of the capitalist system in which we live. How can we act so that people continue to consume and prevent the system from declining? Are we not creating a completely fictitious conception of the individual? A new species of fictitious individuals who would serve as a receptacle for a whole load of injunctions and recommendations, with varying degrees of descriptiveness and normative speech, without showing what the genuine pillars of the conception that we are building really are".

Becoming an actor concerns not only individuals, but also companies. Do they all share the desire for change beyond discourse, or *greenwashing*? Obstacles to the energy transition are still numerous, many legal changes are still necessary, and the preservation of the interest of certain companies still contributes to fueling these obstacles, rather than seeking to change economic models. Fortunately, young people who are the first to be impacted by the success of the energy transition are more and more careful to work in companies that operate and evolve in line with their values relating to their desires to change their habits in order to make the planet more sustainable. The willingness of politicians to create a framework that is conducive to the transition will also be decisive.

On the desire for individuals to cooperate, which appears to be a condition for the success of the energy and societal transition in *La Transition fulgurante* by Pierre Giorgini [GIO 14], the economists Ben Lakhdar, Le Lec and Vaillant call into question the presuppositions of the author surrounding the fact that human beings are inclined to spontaneously adopt cooperative behaviors, and that we live in a world in which information is accurate, with each person supposed to know all of their own possible choices and those of others (theorized in tools such as game theory and blockchain technology [DUR 20, DUR 21, ROB forthcoming, STE 20, STE 21]). Researchers in experimental economics do indeed observe a spontaneous tendency toward cooperation, but this, without a supporting force, tends to decline very quickly. Individuals do not seem to be so much opportunistic or selfish as motivated by reciprocity: the presence of one or more uncooperative individuals is generally enough to demotivate others. The opportunism of some people, which feeds on information asymmetries, affects the behavior of the members of an entire group, and can even destroy this group.

It is not a question of being pessimistic, but of questioning the criteria and conditions essential for the success of the energy and societal transition. Obviously, the technoscientific approach alone is not sufficient to ensure we achieve success. It will be necessary to consider interdisciplinary approaches that integrate engineering sciences, human and social sciences, sociology, economics, ethics, law, etc.

1.4.8. *Small steps to start*

Although slowing down global warming requires strong actions to be implemented very quickly, the importance of the issue should lead everyone to question their ability to act. And yet, it seems as though everyone is implicated in some way and thus can contribute to reducing our carbon footprint by starting with small steps.

Many small individual gestures allow everyone to contribute to reducing greenhouse gas emissions, and therefore to slowing down global warming through travel and the means of transportation used, food, the use and recycling of various appliances, waste, etc. [BAR 20]. It is the beginning of a type of sobriety.

Let us cite a few examples. Transportation is the sector of human activity that produces the most greenhouse gases. It generates around 25%–30% of CO_2 emissions in developed countries, levels which are constantly increasing. Eight percent of greenhouse gas emissions are due to tourism, transport, food, accommodation and traveler purchases. The rise in the standard of living in emerging countries encourages continued growth in world tourism, the impact of which will be only negative on greenhouse gas emissions [WIK 20c].

The bin of an average French person stores more than 400 kg of waste per year and per inhabitant. This corresponds to approximately 200 kg carbon equivalent emitted per person per year [JAN 20b]. Reducing waste and sorting it with a view to recovering it are small gestures that are beginning to occur in everyday life. It is important to note that nature does not generate waste, since everything decomposes and recycles naturally; waste is specifically produced by humans.

It has already been pointed out that livestock farming is a major emitter of greenhouse gases. Eating less meat will therefore help reduce carbon emissions, without having to become a vegetarian.

Many initiatives are emerging among young people and in particular secondary school and university students, for example, to encourage the recycling of used appliances and contribute to developing a local circular economy. Citizen initiatives are also being developed, through setting up repair workshops for used appliances, encouraging carpooling and developing renewable energy sources.

In France, in 2019, carbon emissions of each inhabitant were on average 10 tons of CO_2 per year. Figure 1.13 shows the breakdown of emission sources, with petrol and diesel cars, meat consumption and fossil fuels being the three largest emission sources. To achieve carbon neutrality in 2050, the average emission level should be reduced to around 2 tons of CO_2 per year.

1.4.9. *The need for demonstrators and transdisciplinary approaches*

Implementing the energy and societal transition will entail real-scale experimentation with numerous technological and economic solutions, local societal organizations, etc., running through a series of trials, errors and adjustments. This involves validating the viability of technologies, their efficiency and their impact on the environment, social acceptance, which should not be taken for granted, finding new economic models, designing new legal frameworks, etc. It is also about convincing people in order to limit resistance to change. It is a question of setting examples, of being exemplary…

Science and technology are generally developed in laboratories where the conditions for testing are controlled. This usually follows the development of prototypes for innovations. The complexity and speed of the transition encourage earlier and more systematic testing of emerging solutions in real conditions, in which not everything is under control, particularly when it comes to climatic conditions and human behavior. For example, the technologies will have to be tested, but also ideally designed, in the case of smart buildings, with the users of a building occupying that building as usual.

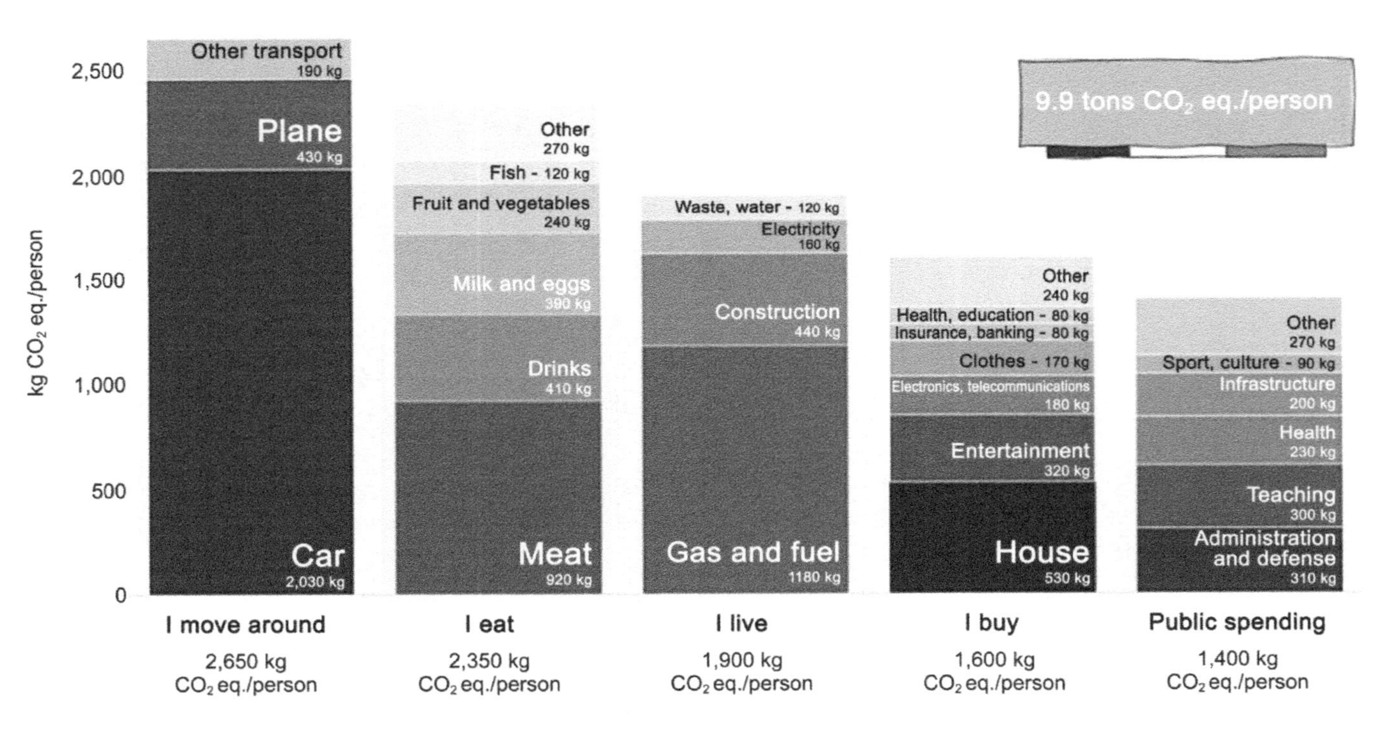

Figure 1.13. *Carbon footprint in France in 2019, evaluated by MyCO2 (https://www.myco2.fr/fr/empreinte-carbone-francaise-moyenne-comment-est-elle-calculee/). For a color version of this figure, see www.iste.co.uk/robyns/smartusers.zip*

A confrontation with the reality on the ground, with what is really happening, requires facing up to many disciplines, both in terms of technologies, electrical, communications for smart grids for example, but also in terms of sociological disciplines in order to assess the acceptability and involvement of economic actors, because new viable models are to be found which are legal, ethical, etc. Interdisciplinarity, even transdisciplinarity, will be necessary for the success of the transition.

The legislator must specify a framework allowing for, or even encouraging, the deployment of demonstrators, wardening off "playgrounds" with "sandboxes", allowing teams the right to experiment for several years.

2

The Transition: Concept or Reality?

2.1. Introduction

When it comes to environmental transition, we may have the feeling that it is about reinventing everything entirely, that we are starting from a blank sheet or that it is a question of building out a new theory.

First, we will highlight that it is necessary and urgent to rebuild a more sustainable development model, that the current model has intrinsic weaknesses and that until now, the answers provided have been generally unsatisfactory. But in the face of this somewhat gloomy landscape, we will also see that there are concrete solutions that have been implemented in many places throughout Europe (Copenhagen, Manchester, Switzerland), and whose initial results seem to be convincing. We will then take a more detailed look at the rev3 approach, created in 2013 in the Nord-Pas-de-Calais region, a region that later became Hauts-de-France when it merged with Picardy.

Bringing all these approaches together will eventually allow us to draw a series of lessons and provide concrete answers to the questions raised due to the urgency of the environmental transition. In particular, it will make it possible to identify the "building blocks" and the cross-functionalities that will prove necessary for any energy and societal transition process.

2.2. The limits of a development model

2.2.1. *An outdated observation*

For several years, and at least within the countries that have a market economy, the feeling that our development model could not last forever has become increasingly present, not to say prevalent. It permeates the reflections raised by multiple think tanks, whether in the scientific, economic, social or philosophical field, with rather heterogeneous responses.

We are confronted by increasingly accurate periodic announcements from the IPCC, alluding to the obvious decline of biodiversity levels (those reading this who are old enough to recall their car windshield covered with insects after a summer evening's drive will not say the opposite), increasingly frequent heatwaves, the disappearance of birdsong and the erosion of our coasts. One would have to be totally removed from everyday life or somewhat unconscious not to realize, or at least feel, that our daily life, the activities within our countries and economic systems will not be able to go on forever if nothing changes.

In the summer of 2021, as in previous years, multiple climatic disasters occurred in various places around the globe, including Europe, which until now had been relatively spared. And if it is not easy for the scientific community to fully grasp the complexity of the issue at hand, the noose tightens a little more each year around a common, undeniable cause, which is that of the behavior of the human species. This behavior is linked specifically but not exclusively to the use of fossil fuels.

But although some of these events have appeared more clearly over the past decade, this is not a recent observation. We just need to go back some 50 years to 1972, when the Meadows report was published in connection with the work of the Club of Rome. This was a remarkable prospective work that studied a set of scenarios and unequivocally demonstrated that humanity could not expect to operate effectively forever using the post-war development model.

For example, one may reflect on the work *Change or Disappear* [GOL 72], which was published in 1972: "it will take much less than a century for our energy consumption to be limited by the thermal tolerance threshold of the ecosphere" (p. 150), and "the rate of carbon dioxide increase that is contained in the atmosphere sits at 0.3% since 1938. The extrapolation of the current trend would give an increase of 18% for the year 2000. According to the EPCE, this could result in a temperature rise of 0.3 degrees on the Earth. A doubling of carbon dioxide would raise the average temperature on the surface of the globe by two degrees" (p. 91) (it should be noted that the title of the work at the origin of these reflections was *Limits to Growth*, and was improperly translated as *Stopping Growth* in the French edition, which is quite different).

Few people outside of small scientific or ecological circles were interested in GHGs back then.

The best known of these gases, carbon dioxide (CO_2), has been monitored for decades at an observatory located 3,700 m above sea level in Hawaii [GLO 21]. When we talk about emissions, we are often talking about a calculation, but here it is indeed a question of an in situ measurement. Here are five values in ppm (parts per million) recorded by this observatory:

– 1960: 316.91 ppm;

– 1972: 327.46 ppm;

– 2000: 369.71 ppm;

– 2019: 411.66 ppm;

– 2020: 414.24 ppm.

Thus, between 1960 and 1972, the average increase was 0.27% per year, it is now (in 2020) 0.62%. The symbolic bar of 400 ppm was crossed in 2015 and by the time this chapter will be read, the 415 ppm threshold will undoubtedly be exceeded. As a reminder, in the pre-industrial era, the value was 280 ppm! The Club of Rome was right…

Twenty years later, in 1992, Rio was carrying out important work under the direction of the United Nations. It is very informative, if a little discouraging, to reread the conclusions and to uncover, for example, the importance that is given to forest management.

In current times, we are rediscovering the importance of carbon sinks, but at the same time our community of nations can only sit back and observe powerlessly the destruction of entire sections of the Amazon, one of the main "green lungs" of the planet. The impact is such that these stretches of land tend to gradually become net carbon emitters, as demonstrated in a recent article published in the journal *Nature* [NAT 21].

Rio was followed by many "COPs" (Conference of Parties) which ended up being failures or for which commitments were often postponed and agreements were difficult to reach.

Significant progress was however made during COP21, still called the "Paris conference", in 2015, but there was no truly binding agreement reached, and 6 years later, few countries have actually implemented their commitments.

The conclusion is therefore clear; it results from several decades of work and observations: our development model creates imbalances on a planetary scale that we are unable to compensate for.

2.2.2. *Having the courage to face reality*

Let us be clear, we know today that it will be very difficult, even highly improbable, to be able to respect agreements that aim to limit global warming to less than 2°C, and possibly to 1.5°C, by the end of the century. But we also know that beyond 2°C (and perhaps even 1.5°C), the world will look different and will carry with it several consequences. Diving deeper into this issue, France, a country whose characteristics we have learned or seen with our eyes, will no longer be the same, whether we are talking of the great cereal plains, the Alpine glaciers, the vineyards that are dotted across the country and even perhaps the organization of life in our cities. Our economy will have to adapt, including but not limited to our tourism industry.

For 50 years, we have come to know the risks that our behaviors pose to us a little more every day, but of course, our industrialized societies have not seriously considered this threat to be a major one, until now.

Awareness of climate change has however, in recent years, been much more widespread than in the past, and the anthropogenic origin of global warming is no longer denied, except by a few stubborn believers who maintain that the Earth is flat, or those who believe in the *moon hoax* (that is to say, that astronauts did not actually make it to the Moon). IPCC's work has been disseminated, made known, and become increasingly precise. In France, the Climate Convention, which was initiated by the Presidency of the Republic, has done a remarkable job by issuing 150 quality proposals, and the High Council for Climate[1] regularly produces highly relevant reports.

At the same time, the reality we face is severe: devastating fires in California and Australia, successive violent hurricanes, extinction of species, temperature peaks in temperate regions, accelerated melting of glaciers, evolution of the coastline, catastrophic floods and an obvious destabilization of the climate in many regions of the globe. So, it seems that humanity is now in the shoes of a patient to whom the doctor has revealed their disease with near certainty, a patient who notices due to multiple symptoms that their disease is progressing but refuses to do what the medical body told them to do. Fundamentally, this means changing their diet and their way of life, a failure to do so being potentially fatal? Everything remains in a state of uncertainty, that consequences "might well" happen. For this patient, even though the worst outcome is imaginable, the date on which it will occur as well as the degree of seriousness it implies remain poorly understood.

1 Available at: https://www.hautconseilclimat.fr/.

The work of scientists, and in particular that of the IPCC, demonstrates without any room for discussion that we have about 4,000 days left (in 2021) to drastically change our lifestyles, the very course of our societies, which is obviously very little time. If we do not do this, it is greatly probable that we will enter a new climatic era, which will be incompatible with the balance of the human species, the animal world and the plant world as we know it.

Origins of this issue have been identified and made known: above all, it centers on an increase in the concentrations of carbon dioxide and methane, which, due to their characteristics (lifespan in the atmosphere, concentration, potential for global warming), contribute most to global warming[2]. The progression of their concentration is directly linked to human activity, in particular to the use of fossil fuels in our daily lifestyles. A simple look at what surrounds us or the course of a single day demonstrates to us their omnipresence.

Recent years are consistently and rightly touted as the warmest in history. A very revealing weakness is being concealed in this statement. When contemplating a past on which we can no longer act, we feel a sense of sadness or resignation, reflecting on the warming rather than projecting ourselves into the future by telling ourselves that if we do not act, each year is very likely to be more temperate than those to come, to finally face reality.

We know that we have to act, and act quickly, in the face of increasingly harsh realities, and we know that time is running out when we ask these haunting questions: How might we change? How might we change the course of things, the relentless march of our societies and our economies? How might we make transitions that disturb or worry people more desirable? Are we going to wait for events that are extremely serious to occur before we take action?

In 2020, a tragedy of planetary proportions came to pass, and though each of us experienced it differently, it is certain that all of us to varying degrees have been marked socially, psychologically or physically by Covid-19. In just a few weeks, the global pandemic brutally reminded us that whatever our dreams, our ambitions, our projects may be, tomorrow does not exist, that our economies and our lives can very quickly switch to an unknown, and they can be subject to harshly limiting adaptations which totally disrupt our daily lives. Perhaps, we should learn from this experience because the same could well be true of climate change, and there, we will not be able to wait for a vaccine to save us.

2 There are other identified GHGs, but they have less impact. The most important, which is rarely mentioned, is water vapor, but its concentration in the atmosphere does not change.

However, this disaster produced the great benefit of showing us that our societies are capable of reacting, sometimes in a very limited period of time, to find solutions, apply them and adapt them continuously. It also brought to light, in a sudden and strong way, a desire that also became a hope, a phrase that has been repeated a thousand times:

> Build a new world that is different from the one before.

It is the same hope that has long existed in the minds of those who understand the reality of climate change, the depletion of resources, the loss of biodiversity, and those who have measured its consequences: the need for a profound change. It is not a question of "starting a revolution" (except perhaps among some people), but of ensuring that future generations, or even present ones, experience a livable and desirable future in a preserved environment on our blue planet which is, until further notice, the only place where humanity can live now and in the coming decades[3].

In the years to come and in the face of increasingly serious climatic events, no one will be able to hide behind the statement "I did not know" or "I was not told". We know what is happening; we have identified the causes and imagined the likely consequences with a high degree of reliability. We have the tools to analyze scenarios, but have we properly measured the degree of urgency? Let us not be (more?) hypocrites; it is time to act.

The time to make a choice has arrived, tomorrow will be too late.

2.2.3. *The intrinsic fragility of our systems*

We are therefore faced with significant challenges, which fundamentally cut across, irrigate and impact all components of our economic and social life.

In front of this, stands the strength, but also the inertia of our economic systems, and in particular industrial ones, which were built after the Second World War and have created a solid base in many countries. At the same time, a complex organization has been built of agri-food resources, mineral resources and energy resources, on a scale that is no longer local, but rather worldwide.

---

3 It alludes here to the utopias that are common among certain players in the space field who imagine, with great persuasiveness, that it will soon be possible to transport tens of thousands of humans to Mars, the only planet that could very possibly become a travel destination. But Mars remains, to our knowledge and within our means, an extremely aggressive environment with a very tenuous atmosphere, which is incompatible with human life [EKS 20].

Democratic governance has taken place in many countries, but it finds itself in competition with authoritarian regimes spread across the globe, including within the G20. In both cases, these are governances that can be described as largely "strong", or at least structured.

We have very elaborate social protection and care systems, territories and urban areas that are structured by increasing numbers of networks and on which we increasingly depend every day. This complex, interconnected, and therefore interdependent world is a beautiful and astonishing construction, and it is clear that this cannot easily be called into question:

– And yet is this world not a colossus with feet of clay?

– Is the inherent complexity of our systems reassuring?

– A few recent examples deserve consideration.

The United States is used to hurricane season, but this has intensified with hurricanes that follow one another and that increase in violence each year. We saw Katrina, then Ike, then Hanna, then Harvey and many others!

Then in February 2021, a winter storm hit Texas. Four million people were left without electricity, the production of natural gas stopped, wind turbines and solar panels stopped working effectively or at all, water treatment systems were paralyzed, and of course without electricity there was no Internet. Imagine what the total absence of electrical energy represents in our societies on a daily basis, with all the implications it would have on daily life, for businesses, public services, etc. The retrospective analysis of this event shows that it is not only the climatic event, the unusual cold which should be put into question. The United States opted to deregulate the energy market, and in particular to provide free sale to the end consumer, which has opened up a perverse game of price fluctuation and investment limitation [MÉR 21].

Another subject that merits reflection: At the start of June 2021, a giant Internet outage occurred which affected major daily newspapers, but also services such as Amazon or PayPal, or institutions such as the websites of the White House or the British government. The origin was rapidly located: it was due to a real-time information dissemination company that faced a technical problem. Confronted with a world that expects to be increasingly interconnected, it is a grain of sand in a gear that ends up paralyzing the whole machine.

The Web has also become an area where crime flourishes, whether purely for profit or a new form of war between countries. At the beginning of July 2021, hackers attacked an American company that supplies software to supermarkets. Hundreds of points of sale were impacted. It was no longer possible to get supplies there, the cash registers no longer worked and essential foodstuffs were thus no longer accessible to consumers.

During 2020, in France, large companies and at least one state agency had been attacked with ransomware. In one case, staff had to be put on sick leave, to allow for time to find a solution (or pay the ransom?) and to put the entire computer system back into service. Just as there will always be thieves, no matter how sophisticated protection or alarm systems may be, there will always be attacks on our networks.

A final example, in a completely different sector: the proliferation of satellites in orbit and the weak regulation of space today lead to paradoxical situations. American entrepreneur Elon Musk's Starlink company plans to put several thousand satellites into orbit, tens of thousands in the long term, in order to provide 24-h Internet access anywhere in the world. This idea, which in itself could be regarded as commendable, means that astronomers who track dangerous asteroids and study the universe cannot see their subjects properly. In the same field, space debris from more and more satellites are endangering the International Space Station and therefore the lives of the astronauts who work there. We have polluted our Earth to a significant degree, and now we are polluting outer space...

We could cite endless examples of fragilities induced by our complex systems, whether in banking systems or in transportation, or even within the supply of basic foodstuffs, even rare resources (where are our main global uranium reserves[4]?).

Weaknesses are partly but not entirely known and will certainly not disappear without practical action. In certain sectors, promising solutions exist: for example, in the field of energy transmission, the complementarity between more local solutions like smart grids, even microgrids, and the very hierarchical network of electricity distribution in France and in Europe is an opportunity that can support the development of distributed renewable energies while providing increased potential for resilience. More generally, relocation, or the return to local supplies or services, can often be presented as a possible solution which is safer, more environmentally friendly and more resilient.

Is it not time to look more seriously at the solutions that need to be implemented to decentralize certain networks, to moderate the power of the markets and to increase our levels of resilience? The current Vice-President (in 2021) of the

4 In Australia (31%), Kazakhstan (12%), Canada (9%) and Russia (8.9%), four countries whose ecological commitment is far from being a model and whose geographical distribution relativizes a sense of independence that is often cited.

European Commission in charge of inter-institutional relations and forecasting[5] and their services in also very interested in this notion of resilience.

But as always, should we not simultaneously return to the behavior of individuals? Should we not also encourage (teach?) everyone to be a little more autonomous, or rather to maintain a capacity for autonomy so that we can deal with the unexpected? We could also relearn how to live a little differently. Before rushing to become a fully interconnected world, even if the law of the market demands it, would it not be better to ask a few fundamental questions about our ability to maintain human life, or even to progress for the common good, and at the right pace? Is it not also the role of our public structures, in Europe, the State or region to help us?

This "multi-fragility" is an entire issue of the transition that seems to us to be hardly or insufficiently addressed today, a subject that could create a link between the issues that depend on local authorities and what emerges from individual behavior. There is no shortage of structures in Europe, particularly associations, that could work harder and increase their contribution. The fragility of the systems we have built is in itself one of its weaknesses which, by "snowball" effect, could one day have serious consequences, even on a planetary scale. What seems worrying to us is not that these weaknesses exist, but rather that they are insufficiently known, analyzed and considered, and several recent crises have clearly shown this.

Whether we accept it or not, our development model, with its qualities and its shortcomings, seems to be getting closer to its limits. It has become endangered on all sides, and it reveals its flaws regularly and does not demonstrate an obvious way out that would allow it to continue developing with confidence and serenity.

So, what do we do?

2.3. Attempts to "repair" the model in the 20th century

2.3.1. *Too strong a temptation*

Faced with these observations, and in light of the accumulating challenges, a search for solutions has very logically existed since the 20th century in a quest which obviously continues today. The difficult question arises when one tries to assess whether these are partial, temporary or permanent solutions. All of these scenarios exist.

---

5 Maroš Šefčovič.

When a devastating tsunami hit the Fukushima plant on March 11, 2011, it was vital to deal with the most urgent matters first, and the Japanese government became busy taking over a private organization (Tepco) which it is fair to say quickly proved to be a complete failure. For a few days, the crisis was such that the strategy to evacuate the four million people living and working in Tokyo was on the table. The repair phase had not yet started; at this point it was more about crisis management. Once this phase had passed, it was impossible to take any action at the heart of the power plant, which needed to be cooled daily; over the years thousands of tons of water accumulated in the tanks. Ten years later, this cooling process continues and we must resign ourselves to an inevitable reality: a million tons of contaminated water will eventually have to be discharged into the Pacific because soon there will be no more room to store it. Though extreme, this is a typical case that is characteristic of a temporary solution which is not sustainable in the long term.

As was presented in the previous chapter, we are confronted with climate change on a planetary scale, whereby today no continent is spared, and part of humanity operates on a model that is no longer sustainable, yet quite logically, many people aspire to this model. The dysfunctions we see are complex because they teem with multiple interconnections.

Let us look at another example: that of the automotive industry, which combines the prominent industries of several countries (China, the United States, Europe, Japan, Korea, etc.) with the need that is firmly rooted in the autonomous habits held by each of us. In France, road transport is the leading source of GHG emissions with 127.7 million tons CO_2 equivalent in 2019 [INS 22]. After driving on gasoline up to that point, in the 1980s, France promoted diesel with tax advantages, particularly favoring company fleets. Diesel was also seen as a way to emit less CO_2 per kilometer traveled compared to gasoline. This was until we realized how harmful the emissions of NOx, SOx and fine particles are, which are significantly more present in diesel engine emissions. Fine particles + CO_2 + NOx + SOx + arrival of the electric vehicle creates a movement that then becomes irreversible. From 2010, the decline of thermal motorization begins and the electric vehicle becomes the holy grail of individual mobility. It is rarely emphasized that the most harmful particles, those said to be PM 2.5 or below, are also emitted by the abrasion of tires and brake pads. Combine this with the problem of battery recycling and one day we will realize that this solution is far from perfect, and that, as many have already pointed out, it is in fact individual mobility that must be rethought.

Once again, we find ourselves more in "repair" mode of the model that is not working than in a questioning phase. Some remark on this, but it seems hard to listen to them, unless you do not want to hear them... On this point, during the first lockdown of the Covid-19 pandemic, it was very revealing that air pollution levels dropped, skies became clearer in metropolitan areas, very few vehicles were on the road in the spring of 2020, and CO_2 emissions fell globally.

Switching from combustion engines to electric motorization is a good thing, but giving to people and companies the means to fundamentally reorganize their mobility would certainly be better. What is happening at the level of individual mobility is thus an attempt to "repair" the model, not to question it. Should we not rethink urban planning and activities that are less dependent on individual mobility, while building locally with all the actors implicated in mobility proposals, in order to come to a set of more suitable solutions? Should we take more notice of the needs linked to different life stages? This approach already exists, particularly in northern European countries.

Taking this path would make it possible to move from an attitude of "repair" to an attitude of real transition. We can see how impactful it may be and how much the temptation to make the least effort or the least disturbance is prevalent.

2.3.2. *The other temptation: the technological answer*

In a few decades, we have gone from the galena set to a complete digitization of information, from mail written with a pen to mail sent with attachments and links. In 1960, 6 million letters a year were written in France, and in 2021 across the world, we will have sent around 320 billion e-mails every day. French car fleets have increased from 13,710,000 on January 1, 1970 to 39,910,000 on January 1, 2019. On January 1, 1950[6], this number only stood at 2,310,000. We have a smartphone in our pocket whose computing capacities are superior to those of the computers of the Apollo capsules which made it possible to go to the Moon. Technology is everywhere and working wonders, whether we speak of medical techniques, data management, predictive modeling or certain branches of artificial intelligence. We were all born into a world which has gone through and is still going through a formidable technological evolution which has been foundational, and which at first sight allows us to live better. Though this is a certainty to an extent, when confronted with some of these figures, there is cause for concern.

Therefore, it is very tempting to say that it is purely through technology that we are going to find technical solutions to the challenges that we have just described, and that tomorrow we will be able to capture carbon, recycle infinitely, dispose of endless amounts of energy, and why not be able to cool our overheated Earth? Some think about this and imagine the most audacious, even crazy solutions: sprinkling

6 Available at: https://fr.wikipedia.org/wiki/Parc_automobile_fran%C3%A7ais#%C3%89volution.

the atmosphere with sulfur derivatives that will reduce solar radiation (China has thus embarked on a process of large-scale cloud seeding [MRM 21]), putting gigantic screens into orbit and even modifying humans so that they are able to adapt to a different climate. And the ultimate "solution" is to leave Earth and go colonize another planet like Mars (see also section 2.1.3) [EKS 20]! This seems like a strange idea at a time when it is precisely the "colonizers" of the past who are strongly decried.

Two questions are thus raised:

– What capacity do we have to measure all the consequences and the collateral effects of the techniques we use? We can return to our dear automobile, which is undoubtedly a technology that has made it possible to make enormous progress, but which we are discovering has also presented serious drawbacks. And yet this is a relatively simple subject.

– On a more philosophical level, could a group of people hold themselves to the right to exercise control over the planet, to try to influence its course, to compensate for a drift for which we are collectively responsible? Does this really make sense? Simply, is it at all decent?

To this first question, the experiences acquired and the observations made since the middle of the 20th century clearly show that we have so far been unable to predict the consequences of our progress and discoveries. Nobody had initially measured the harmful consequences of oil extraction; it was not going to be Henry Ford who would have done it! Today, we are not aware whether the multiplication of transmissions via electromagnetic waves has a harmful effect on our health or not. It took us a long time to realize that refrigerants that we released were destroying the ozone layer which protects us from harmful radiation. The list goes on, spanning the supposed harmfulness of asbestos, to lead, or quite simply, to the misdeeds inherent in the development of peripheral urbanizations and the artificialization of soils. Mankind has always found it very difficult to imagine all the consequences of their choices, quite simply because often this would mean managing a complex system in a predictive way – a system that comprises hundreds or thousands of interactions that are understood poorly or not at all. Therefore, in light of these conditions, having a good time "correcting" the Earth's climate looks like a very dangerous utopia indeed.

As for the second question, it speaks to the ability we might have to place ourselves in an almost authoritarian, if not megalomaniacal, position in relation to our planet. A planet which is around 4.5 billion years old, which has undergone many evolutions and upheavals and on which our existence is a small and quite insignificant event. We know very little of what goes on beneath the surface and what happens at the bottom of the seas. The forces we know how to deploy are ridiculous compared to the telluric forces of the Earth, and yet we are in the process of polluting the thin layer of atmosphere (19 km) which protects us from cosmic rays, and which in its lower part provides the air which sustains us.

Returning to these harsh realities, how could mankind feasibly position itself as the manager of the future of the planet? This would indeed be a very grand pretentiousness, and yet we can be sure that certain “greats” of this world would not reject it if by chance it were to be proposed to them. What level of legitimacy could we give to decisions and risks which would inevitably implicate all of humanity in some way, but which would be taken by a handful of individuals?

A purely technological answer to the challenges that stand before us thus appears neither credible nor realistic, out of proportion to the means at our disposal. We can see it as akin to a headlong rush which, as we know, generally creates more problems than it solves. Of course, techniques will continue to be developed and will certainly enable progress in areas that we cannot even predict today, but there is an urgent need to develop controlled and immediately operational solutions.

Should pragmatism and common sense not make us inclined to question development, to methodically and critically analyze it in order to keep the good and discard the bad? Is this not a job that each of us, and each of our structures, can do right now? Faced with a new project, whatever it is, should we not first ask ourselves what its carbon footprint might be, its ability to create happiness for the women and men it will concern, its real usefulness, even necessity, its impact on non-renewable resources, its recycling capacity or its potential consequences on natural environments?

Obviously, it will be a little more complicated to manage than just calculating the return on investment, but in the end, it will be so much richer and more sustainable.

2.3.3. *The great forgotten issue*

As has already been pointed out, technological accelerations and advances in medicine and artificial intelligence have been considerable since the middle of the 20th century. The speed is such that the world changes fundamentally in the course of a human life, and no one can predict what will in 20 years make up the major part of the environment experienced by a child born today. But at the same time, Geneviève Ferone in her book *2030 le krach ecological* [FER 08] puts it very well: “Today the rarest resource is certainly not the oil or water of tomorrow, it is simply time”.

Time, the great forgotten environmental factor, could well be missed from the equation. Because climate change is already here, it is a process that will not stop, and whatever we do the climate system is experiencing significant inertia.

What should we do over the next 4,000 days?

Certainly not randomly develop one or more of the geo-engineering techniques mentioned above: regardless of their technical or financial feasibility, it would take far too long to make them operational. Probably not, as we have been regularly told for decades, develop an inexhaustible source of energy from nuclear fusion either. Or follow any other hypothesis, which may be entertaining to think about, but which are not in touch with reality: by the time they are implemented it will be too late. So is there still hope? Yes, perhaps, and we propose here to concretize it around three axes of thought:

– the first is to become aware, individually and collectively, at all levels, of the time limits before us in order to make more responsible choices now;

– the second is to act with pragmatism and rely on known and proven techniques and methods. This does not mean denying the importance of innovation, quite the contrary;

– the third is to develop successful experiences much more broadly and quickly; there is no shortage of them!

The paradox which is more than apparent is that in a world where everything moves faster and faster, it would be beneficial to slow down in order to reflect, think better and decide better, all the while moving much faster in the transition from our societies, because time is running out for us. The exercise is difficult but enriching, and it is particularly on the radar of European institutions.

2.4. Cities and territories in transition

The three approaches mentioned above have fortunately been applied in many places in Europe. So much so that presenting an exhaustive vision of what has been done or in progress would require several books. As such we do not claim to be exhaustive here, as the experiences are multiple, rich and diverse. We will therefore limit ourselves to a few examples that we find particularly significant:

– the Copenhagen approach;

– the Manchester project;

– the Swiss project of the 2000 watt society.

2.4.1.2. *European Green Cities Index*

The second study, which dates back to 2009, was conducted by the Economist Intelligence Unit, and sponsored by Siemens [ECO 09]. The figures quoted below are taken from the report on this study.

This work analyzes and qualifies the performance of 30 European capitals, and around 30 indicators are grouped according to the following eight criteria:

– CO_2;

– energy;

– buildings;

– transport;

– water;

– waste;

– air quality;

– environmental governance.

These are the main conclusions:

– The Nordic cities are leading the pack: Copenhagen, Stockholm and Oslo take the first three places. They are followed by Vienna and Amsterdam. At the other end of the ranking, we find Bucharest, Sofia and Kyiv.

– There is a strong correlation between the level of wealth – measured by per capita GDP – and the overall ranking. It is not always significantly differentiated on individual criteria.

– The size of the city does not seem to be a major criterion, even if the results of the smallest cities (less than one million inhabitants) are generally better.

– The involvement of civil society and its residents, through associations or professional or religious groups, plays a positive and important role in the ranking.

– The impact of joining the European Union is significant, obviously through the redistribution of financial resources that it offers, but also through the effectiveness of awareness campaigns, through the major environmental programs that it promotes and by applying European directives in Member States.

Of course, historical pasts have an impact on the performance of cities, in particular those which have lived for a long time under communist regimes, for which the maintenance and quality of infrastructures have not obeyed the same rules as those of Western countries for a long time.

What seems to us to be particularly instructive in this report is the great variety, if not heterogeneity, of the results uncovered for each criterion. To understand this better, we are going to review some of these criteria, starting of course with CO_2 emissions per year and per capita (in direct emissions). At the very bottom of the scale, we find Dublin with 9.72 tons, and at the very top Oslo with 2.19 tons.

If we look at energy consumption, it is still in Dublin where consumption per inhabitant is at its highest with 156.46 GJ, but it is Belgrade which is best placed with 41.07 GJ (with the exception of Istanbul, for which the value is estimated) while Oslo is at 94.78 GJ.

The final indicator is the percentage of renewable energies in the energy consumed by the city. This is also where we see the impact of emission reduction policies launched quite some time ago. Oslo is at 64.8% immediately followed by Stockholm with 20.08%, while many cities have very low percentages, below 1%.

A few takeaways are as follows:

– out of the 30 cities, many could do better, if they wanted to do it with a better control of consumption. Ambitious goals are achievable now and are not utopian;

– programs that aim to achieve carbon neutrality or that set the ambition to reduce GHG emissions obtain tangible results after several years;

– energy is a real subject relating to urban progress which today relies on proven techniques and therefore deserves to be considered as a priority.

On the subject of buildings, the differences are no less great. Berlin and Stockholm share first place in an aggregate indicator based on three indicators that are all linked with energy. Once again, the lesson here is that anteriority pays off; Berlin launched an ambitious urban renewal program in the 1990s. In 2009, two-thirds of the park in the former East Berlin had been renovated. In Stockholm, regulations have evolved for a long time by imposing very high standards compared to the standards set in other European countries.

Considerations of the same order relate to water, waste management and transport.

The first lesson of this study is that positive performance can exist everywhere; it can be obtained with varying levels of ease depending on the level of wealth, geographical location and historical criteria, but it is very encouraging to see that it is possible, when placed alongside the European Union's 2030 objectives (*Fit for 55* package) and carbon neutrality in 2050.

One of the keys is to have an ambitious, shared, solid plan, and to afford it strong governance. We will see below the remarkable work of Copenhagen on this subject. Five years later, 10 years later, sometimes more, the results are there.

It remains true that at the crossroads of all these experiences, the themes to be addressed can be perfectly identified today by what we will now call the "building blocks".

We can therefore draw a message of hope from this work, a conviction that in the complex web of interactions between all of these possibilities, at the crossroads of all these techniques and infrastructures on which the city is based, there is potential for a certain level of progress and development, a response to the present challenges which must and can be adapted to each situation and which require the participation and commitment of all the inhabitants, as much as an informed governance of territorial structures.

2.4.2. *European cities and territories*

Fortunately, these solutions have already been implemented and materialized in many territories. What lessons can we draw from this? This is what we are going to try to highlight with two cities in Europe and a country which have started to publish the results of their procedures.

2.4.2.1. *The beautiful ambition of Copenhagen*

The capital city of Copenhagen shouts loud and clear about their ambition to reach carbon neutrality: in 2012 it announced a strong, clear and emblematic ambition to be "carbon zero" by 2025 [MCC 20]. This is one of the rare territories of this size within Europe where such a mission has been promoted for many years (others have since embarked on it). And it is all the more interesting to see what is being done there since recent headlines seem to speak to the success of the process.

The Copenhagen agenda is structured around four main areas:

– energy consumption;

– energy production;

– mobility;

– initiatives from municipal governments.

It is always done with the backdrop of reducing GHG emissions, which is the primary objective that will guide words and actions.

But the Copenhagen plan had actually been initiated in 2009, meaning that it is over a period of 16 years that the approach of this city has spanned, which from the start had significant advantages on the subject. More recently, an interim report redefined the 2017–2020 roadmap.

Figure 2.1, which has been taken from this report, shows the evolution of CO_2 emissions predicted in Copenhagen.

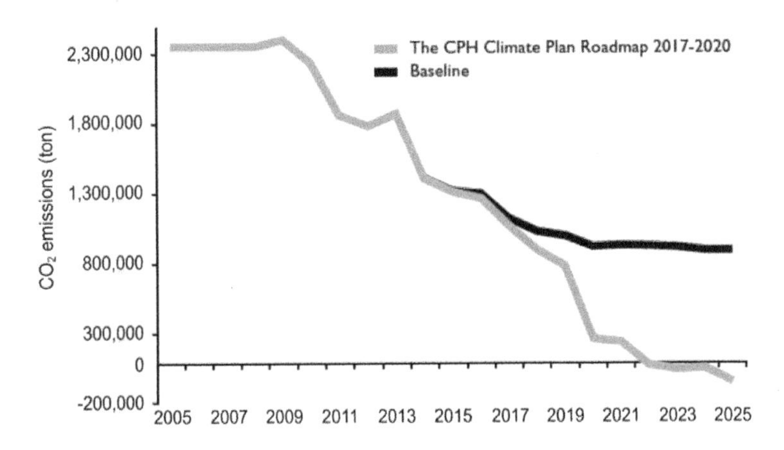

Figure 2.1. *Evolution of CO_2 emissions in the city of Copenhagen between 2005 and 2025 [MCC 20]. For a color version of this figure, see www.iste.co.uk/robyns/smartusers.zip*

Analyzing this curve uncovers two key takeaways:

– from the start, during the period from 2009 to 2015, measures have been effective, as emissions are falling steadily;

– however, from 2016, the planned measures prove to be insufficient (black curve), and in 2017, new measures were taken which should make it possible to achieve carbon neutrality in 2025 (green curve).

Here are the main actions carried out between 2013 and 2016 for a budget of 105 million euros:

– centralized heat and electricity production (cogeneration) gradually shifted from coal to biomass;

– garbage collection is done mostly by gas-powered trucks;

– 64% of municipal vehicles are either electric or hydrogen;

– 20,000 urban lighting lamps were replaced by LEDs with an estimated energy saving of 57%;

– new buildings must meet 2020 regulations and ambitious renovations are underway;

– cycle lanes and footpaths have been developed to further facilitate the development of soft mobility.

The following roadmap covers the period 2017–2020, and is based on four points that can be summarized as follows:

– continued reduction of heat and electricity consumption;

– the development of carbon-neutral urban heating, together with the production of renewable electricity which is higher than the city's consumption levels, all with a view to make energy sources more flexible;

– very ambitious objectives in terms of mobility, with 75% of urban journeys made on foot, by bicycle or by public transport, but also the arrival of new types of motorization (electric and hydrogen);

– a strong desire to set an example on the part of the municipal administration in terms of buildings, travel and purchasing policy.

Beyond the technical characteristics at play, what is remarkable in this plan is on the one hand the method followed, and on the other hand the fact that all the projects are evaluated on the basis of measuring CO_2 emissions. A *Guardian* article published in 2019 [THE 19] makes a very positive assessment of the plan, of its progress, and is quite explicit on the context in which this plan is carried out, highlighting its broad acceptance by the inhabitants.

In 2025, it is therefore more than likely that at least one European capital will have achieved carbon neutrality.

2.4.2.2. *The Manchester plan*

Manchester has created a much more recent approach which is still in its first phase. This plan, which is called the Manchester Zero Carbon Framework 2020–2038, spans a period of 18 years.

Manchester offers an extremely structured approach, which emphasizes all of the opportunities and benefits of all types that the plan will bring to both residents and economic activities. The proclaimed objective is to reduce GHG emissions by 50% in 2028 compared to 2022 and to achieve carbon neutrality in 2038. This is based on the threefold desire to take advantage of the best scientific approaches, to respect the commitments of the Paris agreements and to involve residents and businesses.

In the first objectives announced, we can find subjects that are becoming “classic”:

– energy efficiency of municipal buildings;

– planting trees in urban areas;

– development of solar and wind energy on a large scale;

– implementation of electric dump trucks for waste collection;

– urban lighting;

– replacement of polluting vehicles by electric vehicles;

– walking, cycling and public transport becoming the main means of urban transportation.

And in parallel, several proclamations:

– expected health benefits and therefore savings in medical care;

– energy savings;

– a new dynamic of job creation and activities around zero carbon;

– reduced food waste;

– a reduction in traffic jams.

Sixty organizations have committed to the plan, and each has submitted its own emissions reduction plan. The principle adopted is that of the “carbon budget”. Manchester first defined its 2018–2100 carbon budget, that is to say, its right to emit for the rest of the century, based on national strategies. Three multi-year durations – 2018–2022, 2023–2027 and 2028–2037 – were then defined. Each period is assigned a carbon budget, and this decreases over time. Direct emissions in 2018 were around 2.3 Mt/year. Over the first 5 years, they plan to lower this value to an average of 1.39 Mt/year, over the next five years to 0.72 Mt/year and over the next 10 years to 0.30 Mt/year. This may seem ambitious, but it is what we must confront if we want to face the question of GHG emissions with lucidity and honesty in the decade to come.

2.4.2.3. *The Swiss project for a 2,000 watt society*

This project was initiated before 2006 due to specific requests from the city of Zurich. The term “2,000 watt society” is unusual and deserves to be explained. It results from the observation that at the beginning of the 21st century, daily power requirements were around 6,500 W per person and corresponded in Switzerland to CO_2 emissions of 9 tons per person per year.

From the start of this project, two universities (EPF Zurich and EPF Lausanne) and four research institutes were joined together, and dialogue was opened up with the general public. The work carried out, particularly by the university sector, shows that a sustainable future, which would make it possible to contain the rise in temperatures below 2°C, requires one to converge toward a value of 2,000 W and limit emissions to 1 ton of CO_2 per person, per year.

What distinguishes the approach of the Swiss society from the outset is that it immediately considers the use of materials and also talks about efficiency in this regard. Basically, it establishes a very strong link between energy and materials, then between materials and consumption. The term used is "for a sustainable and equitable exploitation of resources". It should be noted that the attention given 15 years ago to materials was a precursor, as it reappears today with great relevance in France within several transition scenarios (Ademe, NégaWatt, see section 1.4) [ADE 21, NÉG 21].

In summary, the 2,000 watt society project is based on five areas of action:

– housing;

– mobility;

– food;

– consumption;

– infrastructure.

There are very clear findings and objectives. Housing, which is estimated on average at 1,800 W, would decrease to 500 W. By going back to the calculation over 8,760 h (1 year) and an average surface area of 75 m^2, this corresponds to a transition from 210 kWh/m^2/year to 58 kWh/m^2/year, we thus find objectives close to those held by France. But it is important to note that this is an objective for existing buildings and not just for new ones.

It is more difficult to measure the effort in terms of mobility, which is rarely expressed in the power required. If the project presents a switch from 1,700 to 450 W, it also mentions an increased use of soft transport and public transport, as well as an objective of "less than 9,000 km traveled per year with an economical vehicle".

We will not be surprised to learn that we must also limit our meat consumption. However, it is more original to consider the same level of ambition as the first four subjects in a fifth subject, which surrounds a call for the authorities to drastically increase the efficiency of infrastructure and public goods. From the start of the program, the focus is on consumption (clothing, furniture, electronic devices) with a clear message put forward: "make what you have last".

What is very striking when reading this project which, let us recall, was initiated at the beginning of the century, is that it asks the right questions, or what we have called the "building blocks" which we will end up finding everywhere in one form or another.

Where is the project at today (in 2019–2020)? It still lives and continues to exist, now holding onto three objectives:

– energy efficiency (2,000 W of primary energy per citizen per day);

– climate neutrality (zero GHG emissions related to energy consumption);

– durability (100% of renewable energies).

The monitoring that has been carried out shows that energy consumption is decreasing, as are GHG emissions. Large cities are involved (Zurich, Lucerne, Aarau, Winterthur, etc.), the concept of smart city is spreading, and it is understood and presented in a very open way and is not limited to the technological space.

In summary, the major topics to address were identified first, before focusing on energy, appropriation and implementation. It was after this that we saw the first positive results come to the fore. All further information is available on the project website[7].

2.4.3. *Rev3, the Hauts-de-France project*

It is 2012, at the World Forum for a responsible project economy[8], in Lille (France). Jeremy Rifkin, an internationally renowned prospectivist, in front of a full house, was able to present his vision for the Third Industrial Revolution. It was the meeting of three men, Daniel Percheron, regional president; Philippe Vasseur, president of the regional chamber of commerce and industry; and Jeremy Rifkin, who will give birth to a major transition project, which crystallized a year later in the form of what was called the Master Plan for the Nord-Pas-de-Calais region, a

7 Available at: www.2000watt.ch.

8 The World Forum for a Responsible Economy is an international event held in Lille (France). It lasts 2–3 days, during which renowned speakers are invited, and round tables and exchanges are organized on the theme of CSR. See https://www.responsible-economy.org.

region in northern France which at the time had four million inhabitants. This project was born under the name of Third Industrial Revolution on October 25, 2013, once again at the World Forum, for a territory which was back then Nord-Pas-de-Calais. It has since become "rev3" and has extended to the entire new region of Hauts-de-France.

It is not the purpose of this book to describe in detail the progress that the region has made, nor the governance that has been put in place. Lots of information of this genre can be found on the project website[9].

It seems more interesting to us to indicate here the main aspects of the shared vision, its evolution over the years, and the concrete results that this approach has achieved.

In 2013, the work of reflection and formalization was carried out on five pillars:

– The development of distributed renewable energies. This qualifier is important. Rifkin always says *those in my backyard*, particularly the sun and the wind, two sources that "do not send a bill", exist all over the world and remain for a long time. And in the same way, the twilight of fossil fuels is an essential condition for reducing CO_2 emissions.

– Buildings: new or renovated, they must consume very little energy and, when equipped with renewable energies, they will become producers of energy. Bringing built stock to such a level of performance would also be a huge contribution to jobs that cannot be automated or relocated.

– Mobility: for obvious reasons, we must abandon the use of thermal engines as quickly as possible, switching to electricity and hydrogen, but also evolving toward more soft modes of transport, cycling and walking, whenever possible.

– Energy networks: as a direct consequence of what was written above, but also due to technological developments, it is necessary to gradually move from current, highly centralized schemes to meshed ones, at least partially and locally. These networks are equipped with efficient piloting; we are talking about smart grids or intelligent networks.

– Energy storage: as a correlative consequence of developing renewable energies, it is necessary to be able to store large amounts of energy, as wind power, photovoltaics and hydraulics along the water [ROB 15, ROB 19, ROB 21] have the weak point that they cannot be easily controlled. Different techniques are now available, while others are under development.

9 Available at: www.rev3.fr.

It will be noted that these five pillars are not independent, but strongly linked which leads one to make binding choices. This becomes even more apparent in three cross-cutting areas, which are the following:

– Energy efficiency: significant consumption savings are possible today, in particularly but not exclusively in the field of buildings. Industry in particular is a sector that should be considered depending on the region.

– The development of the circular economy: though recycling is one of the major components of this, eco-design, reuse and repair are all tracks that must be followed and developed, in the face of shortages of materials, or problems sometimes posed by locating or extracting them.

– The development of the functional economy: sharing rather than owning an underused good, developing mobility services, for example.

The reflection on the future of energy in Nord-Pas-de-Calais has resulted in the graph shown in Figure 2.2, which is a very simple yet instructive way to illustrate how the production of renewable energy could meet consumption by the year 2050, in a future where 100% of energy would be renewable in nature (see section 1.4.3).

A year of work has given rise to a set of strong, varied and coherent proposals, but also to a shared vision. The main feature of this work is that it was carried out jointly by political forces (regions, departments, urban communities) and economic forces (chambers of commerce, companies, etc.), and there were also stakeholders from the world of associations and the academic world.

The strength of this approach came from its definition of a narrative, a possible story for a region which was then projecting itself into a desirable future, which was not however a utopia. It was a story of hope, of development potential in many areas, and already back in 2013 an answer to the pressing questions that the evolution of climate now reminds us of on an almost daily basis.

It is the relevance of this narrative, the characteristics of shared governance and the support mechanisms that has been put in place that have ensured that this great ambition has survived over the years. More specifically, it survived the merging of regions, which came to pass at the end of 2015 when Picardie and Nord-Pas-de-Calais were brought together to form a large region which was to be called Hauts-de-France in 2016. The merger of the two regions could have stopped rev3. However, on the contrary, the new executive has from the start wanted to extend it to the whole greater region. Although of a different political orientation from that of the previous executive, he has taken over and carried rev3 without qualms, making it an assertive tool for progress and the creation of labor.

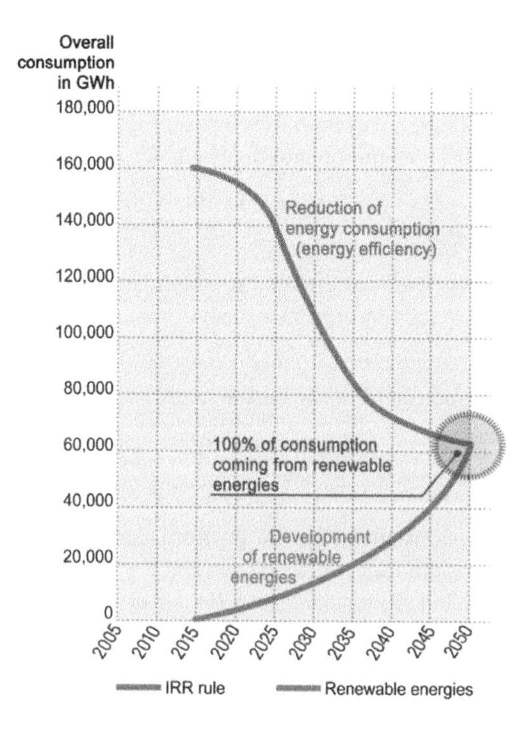

Figure 2.2. *Evolution of the consumption and production of energy in Nord-Pas-de-Calais*[10]*. For a color version of this figure, see www.iste.co.uk/robyns/smartusers.zip*

Many project leaders, who from the start mobilized themselves by integrating the approach, have also made this their commitment. At the end of 2021, there were approximately 1,300 projects of all kinds recognized in the process, and according to the region nearly 160,000 jobs have been created. This was made possible because of the economic world which knew how to mobilize, but also because the political power constantly supported rev3. The rev3 mission which brought together employees from the region and the regional chamber of commerce and industry (CCIR) has made it possible to join forces and maintain a seamless dynamic. Once again, readers can find all further information on the rev3 site. After 8 years running the project, we are able to draw some lessons from it by considering it from the perspective of a transition project, and to identify its strengths, weaknesses, opportunities and threats.

10 Available at: www.rev3.fr.

What are its strengths?

– They reside in the beautiful dynamic that has been put in place and has never weakened, with collaboration carried out by the political and economic forces, both public and private, distributed throughout the Hauts-de-France territory.

– They are also related to the academic world, with close proximity to the many regional centers of excellence.

– They are also in the wide variety of areas implicated and impacted, whether we talk of construction, logistics, energy mix, heavy industry, agriculture, training and many other areas.

– Finally, they can be found in the narrative created and in the connection kept with Jeremy Rifkin.

Its weaknesses? We prefer to speak here of points to watch out for:

– The first is undeniably the ability to go from demonstration projects to large scale ones, to massification. Renovating 10 houses in a very efficient way is proving to be full of lessons, but doing the same for several thousand is a completely different matter, and the same can be said of the development of renewable energies, the evolution of infrastructures, the development of the circular economy, etc.

– The second point is linked to the previous one: it is difficult to anticipate the new resources that massification will require, regardless of the areas concerned, and therefore there will be the potential for difficulties when it comes to implementing projects.

– The third point to watch out for is crucial; it lies in our ability to react quickly at a time when climate change requires us to do so. Our decision-making methods are often too slow, too cumbersome or unsuitable for the specific situation (see also section 2.2.3).

Opportunities? There are many:

– The first is incontestably offered to us by the European Green Deal. This refers explicitly to the Third Industrial Revolution and many of its themes have already been implemented in Hauts-de-France [COM 21].

– The second is to make it even more of a pressing subject in the region, even closer to its inhabitants, because although the economic world is directly involved, rev3 is still little known to the general public and all the social and societal implications of rev3 need to be developed.

– A third opportunity is to accentuate the link between an environmental transition and a digital transition, as this link is very present in the objectives of the European Commission. This is one of the points of development that rev3 can still work on. It appeared in the initial stages of the project and has asserted itself over the years as a real opportunity.

– A fourth opportunity – which once was a weakness, but which is on the way to becoming a great strength – is the commitment shown by the industrial world. Initially not very present, even sometimes hostile to the ideas presented in rev3, the world of industry is today more than just present, it is creating actions and is happy to do so. We can cite the remarkable dynamic that has developed in the Dunkirk area as an example, relating to the CO_2 decarbonization program, industries and territories in which the ArcelorMittal group is involved.

Are there any threats? Of course, as with any large project:

– The first is perhaps the ubiquitous difficulty relating to working cross-functionally. As we said earlier, the major projects that drive rev3 forward are essentially cross-functional and require the ability to federate and bring people together, which is not always easy in our country.

– The second is excessive regulation, which can constitute a brake on progress and innovation. This is a well-known threat, already encountered in a Jacobin country (or centralized state).

– The last is the presence of interests that are monopolistic in nature, the force of which could slow down the project or direct the implementation of new structures or processes to be more advantageous to them than to the wider approach.

One thing is certain: rev3, for those who have experienced it, is a great and exciting adventure, which brings progress and hope to the region. Like any human adventure, it is sometimes imperfect, but it harbors immense qualities and is rich in learnings that can be brought to bear today.

2.4.4. *Some lessons learned*

As a starting point, it should be recalled that these examples mainly concern urban society. It is clear that the rural world has, and will continue to have, its fair share of contributions to this necessary transition. Rather than being an oversight on our part, we believe that it is a field that, though not explored here, deserves attention.

We could extend our analysis to other cities and territories in Europe or to many other countries of the world, but the examples which have just been cited are already sufficiently rich. Here is a first list:

– A strong dynamic supported by well-established governance. Our behaviors, our economy and our humanity have constantly evolved over the centuries, along with our cities and territories, but within a dynamic that could be described as induced, natural or unconstrained, except during times of war. Today, the pressure that is now being exerted by climate change forces us to rethink our models and our behaviors, in order to make complex and difficult choices. And it is obvious that each project presented here has its imperfections and shortcomings, and yet it tries to answer the questions which have become fundamental, and is organized "in working order" to be able to get there, with the support of project governance and a guarantee of assured dynamics.

– The projects converge. Conducted independently of each other, in different countries, the projects are for the most part saying the same thing. This is what will allow us to identify what we call the "building blocks" of a transition (see section 2.5). Throughout all the projects, the identification of these blocks comes from a reflection, which is originally scientific in nature, and is based on measured and/or observed facts, rather of a technical nature. However, due to its nature, it very quickly extends to the field of the human sciences.

– Coordination between political and economic structures is manifested differently depending on the project, yet it is present everywhere. This coordination which is initiated from the very start of the projects is important in terms of what could be called the performance of the projects, and their ability to achieve concrete results when it comes to commitment. We can see that this come to light most clearly and openly in rev3, and the Hauts-de-France region has effectively demonstrated over the years that this was one of the strengths of the project, and one of its main points of support.

– The relevance of the notion of territory. The multiplicity of components within a global transition plan, the implementation of human, financial and material resources, is not always within reach for a small town (which does not mean that they cannot do anything at all!). Conversely, at the national level, disparities between regions make it difficult to align on and implement a single transition project. From the experiences that have been analyzed, it seems to us that the notion of a metropolitan area represents the lower end, and the European region perhaps sits at the upper limit. Between these two limits, there are groups which have the means to establish a structured plan, to carry it out, to associate all the academic components that will contribute to research and innovation, and to mobilize sufficient economic processes. The notion of critical size is essential. The Copenhagen agglomeration has 1.3 million inhabitants, Switzerland is a country of 8.6 million inhabitants and Hauts-de-France is a region of 6 million inhabitants...

– The project must be shared. That is to say co-constructed, to integrate all the social and societal dimensions that it entails. Alongside political, economic and academic forces, the participation of social bodies, associations, but also of young people, in an era where the speed of change is rapid, is essential, so that the choices made fully integrate with the needs, aspirations and reflections of these age groups.

As a sensitive subject, the probable prospect of the profound changes that will be required on the mentalities and behaviors of individuals will have to be understood, accepted and integrated, with the hope of reaching a renewed stability in the medium term on many levels (social, economic, climatic, etc.).

– The strength of the narrative. This is a rarer component, which seems in our opinion to be lacking in many transition projects. This component can be seen in the title of the Swiss project – to speak of a society at 2,000 watts is to raise certain questions, but it is also to lay the foundations for a statement: it will be the main subject of a society or it will not be, and it is a matter of energy. The terms used are unifying. Due to the fact that it was within this unifying configuration that the project came about, in Nord-Pas-de-Calais the narrative carried us; this is one of the great contributions from Jeremy Rifkin. Throughout 2013, the narrative was built and seamlessly woven around the usual components expressed by Jeremy Rifkin, but it also managed to create a real sense of connection to the specific characteristics of the region. In the end, the narrative of a desirable future can be told, it can unfold logically and produce a sequence of words that give it its strength. The narrative is based on what already exists and has been carefully analyzed, and offers us a vision that provides the answers to many of our questions. It opens up a space for implementation and invariably propels actors toward new responsibilities.

– The "large forgotten issue" of section 2.3.3 is now essential. If Copenhagen achieves carbon neutrality as planned in 2025, it will have taken 16 years of constant work, carried out on the scale of a metropolis, to get there. Other examples, which are not detailed here (Thisted, Sønderborg, Oslo, etc.), display an ambition of between 15 and 20 years, but never less than 15 years. This must be compared with the *Fit for 55* package announced in the summer of 2021 by the European Commission, which sets a target of carbon neutrality by 2050, that is to say in 50 years, and a reduction in GHG emissions of 55% in 2030 compared to 1990 [MIN 21], in other words in nine years. Europe has reduced its emissions by 23% in 28 years (between 1990 and 2018); therefore, the objective to reduce the remaining 32% by 2030, in nine years, represents an effort three and a half times greater.

If immediacy is essential and the time for procrastination is behind us, what are we going to do tomorrow?

The time for choices is upon us.

2.5. Create a systemic approach

In the Hauts-de-France project, the emergence of a need for a global, cross-cutting approach quickly appeared, because of the reflections of the eight working groups that had been set up. It was by working on their own topics that several groups mentioned the need to discuss issues with others. The group on buildings needed to take advantage of renewable energies and the circular economy, energy to networks and storage, etc. The difficulty is that our organizations are vertical and most often operate in silos, with "walls" to break down and "preserves" from which we need to break free. Even during the drafting of the *Master Plan*, we had to fight against this handicap.

At the same time, the conviction already stated to distinguish the "building blocks" of a transition only reinforces this need to define the global system in which they can be integrated. It is the construction of this global approach which will be one of the characteristics of the systemic approach.

2.5.1. *Building blocks*

The analysis of the projects presented in this chapter (and other projects not mentioned) leads us to propose what could be the "building blocks" of a transition approach today. According to territory and depending on the main sources of emissions and the origin of these sources (industrial, heating, vehicles, etc.), their importance will of course be modulated, or even a few other specific "blocks" will appear, but there is a common base. These "blocks" will also be linked and made coherent, because of transversalities that we can find in the different projects.

So, let us first outline a definition of the building blocks, without order of priority or importance:

– *Energy and the development of renewable energies*: this is an invariant found in all plans. Energy production and consumption are major sources of emissions, and the development of renewable energies is found in all scenarios. Very logically, we will integrate the need to develop storage capacities when it comes to electrical energy, or even new networks capable of managing localized individual production as effectively as possible (smart grids).

– *Mobility and logistics*: whether for individual travel or the transportation of goods, road and air transport are major sources of CO_2 emissions accompanied by various forms of pollution. Their impact is considerable, and many solutions exist, often starting with the moderation or optimization of uses. However, we must not minimize the deep work that is still necessary to carry out on individual mobility, whether it is done for professional or personal reasons, it is a project of huge proportions.

– *Renovation of buildings*: most buildings in Europe are not very efficient, because even though the current regulations oblige them to build in a very efficient way, the existing stock renewal is only at 1%–2% per year and the modes of heating systems remain high GHG emitters, including in France. This therefore implies that we must make an unequivocal effort to efficiently renovate buildings[11], as the *Green Deal* has also clearly underlined.

– *Digital technology*: less present, or sometimes absent, in the oldest transition plans, digital appears both as a solution, partial or revolutionary depending on the case, but also as an area requiring a high level of energy, and it is therefore a great GHG emitter. Again, solutions are being developed, but they must be monitored in a "balanced" way, looking at the reduction of emissions that come from the use of digital tools and the increase in emissions linked to the operation of these tools.

– *Food and its corollary agriculture*: this area is well recognized as one to consider in a transition plan. Whether it concerns production and distribution methods, the consumption of animal proteins, transport or even food waste, all these subjects deserve thought, measures and actions.

The first five building blocks are therefore:

– the development of renewable energies;

– the evolution of mobility and logistics;

– the renovation of buildings;

– the integration of digital technology;

– food and agriculture.

These are five blocks with which we often associate industrial processes (depending on the territory), and which often consume a lot of matter and energy.

Obviously, these foundational blocks cannot remain independent from each other within a transition project, especially if it offers a narrative, a global vision.

What then are the main transversalities that are necessary for a transition?

2.5.2. *The unavoidable transversalities*

The first of the transversalities is energy efficiency, the ability to produce, to live, to inhabit and to move around in the same way while using less energy. This is an imperative that is essential in the context of a gradual and essential abandonment of fossil fuels and their replacement by renewable energies. Energy efficiency has a

---

11 Which, in France, means being at least at the level of the BBC renovation label (a low consumption building), which corresponds to conventional consumption that is less than or equal to 80 kWh/m2/year, corrected by regional factors.

corollary which is energy sobriety, which is the subject of many reflections (see section 1.4.2). These are two different though related words.

Accustomed to inexpensive energy sources, will we be able to change our behavior to make more reasoned, more moderate use of it, or even face up to certain limits? This is even less obvious. Since the 1970s, our societies have been built on multiple abundances, whether in the field of energy, mining resources, the ability to come and go, etc. Low-cost air travel is one of the most remarkable examples of this, and also one of the most polemic. However, in terms of energy as well as in the use of materials[12], or even in terms of movement, sobriety represents a potential opportunity that can no longer be ignored and deserves to be integrated into reflections and deeds. Even if this subject concerns the social field of the transition above all, it can be enhanced by the progress being made elsewhere, for example, in the field of data and its use. Debate will probably be called for in the years to come, once we come up against the limits of the current model (see section 2.2).

The second transversality is the development of a much more circular economy. Theorized for a long time, but regarded with little interest until recently, the circular economy is emerging as a path to progress in all areas of society. Whether through recycling, reuse, renovation or repair, this mode of operation has great advantages as soon as we take into account the life cycle analysis of the equipment, an object or a structure. This form of economy brings the time dimension into the consumption of goods. It is an essential element from when the asset is first designed, whether it is a tool, a vehicle or a building. Its corollary is the functional economy, which aims to replace the possession of a good by the provision of a service. Do I need to own a vehicle, a tool, or more generally a material good or is it enough for me to have one available when I need it? These are two economic models that are very much linked to behavior, to the social acceptance of acting differently. Which brings us to a third transversality.

This third transversality is a new social model. This concerns everything that will affect people, society, in a sharing economy which to this day is still very far from being accessible to everyone. It concerns the development of new behaviors that result from the considerable impact of digital technology in our daily lives for example, but also from the possibility of becoming an individual energy producer, the possibility of training online (MOOC, Massive Open Online Course) or the ability to co-develop proposals. It can be participatory but also the advent of new forms of precariousness. It is also about how the world of education and training is adapted. The foundation of our society, which is impacted by major upheavals, must reinvent itself so as not to sink into chaos, but perhaps be reborn a little differently,

12 The United Nations Environment Program (UNEP) estimates that the production of materials is responsible for 23% of global GHG emissions.

in a form that is more resilient, less violent, after having revisited the values on which it relies.

In summary:

– *efficiency and sobriety,* which can be translated as “less but better”;

– *circular economy and functional economy* to reinvent a more responsible economy integrating the time dimension;

– *new social models*, new behaviors, which are more inclusive and which demonstrate reaffirmed values.

The base of these transversalities therefore makes it possible to coherently advance the development of the building blocks in a global and systemic approach to a transition which could be seen to be ambitious, in particular in terms of reducing GHG emissions.

We see that the territories involved in a real transition are those that roughly follow such a pattern, and as was mentioned earlier, following this they begin to see results and immense benefits.

2.5.3. *Buildings, one of the hearts of the transition*

There are several themes at the heart of the environmental transition, which we know deserve a global, systemic, cross-cutting approach, and whose impact is major in contributing to the aforementioned objectives, in particular those concerning the year 2030.

Of course, mobility, digital technology, buildings, industrial activity and agriculture form part of this. These are all subjects linked to the social, environmental and economic components, and for which the reduction of GHG emissions is a primary, unavoidable, even urgent objective. They appear in the *Fit for 55* package and all align with the title of the present work, which talks about intelligence, users and energy and social transition.

Let us state here once again our conviction that this transition cannot be carried out solely by a political power, whether local, national or European. It will only happen if it is the fruit of a common will of the elected powers, of the economic world and of each of us in a common and shared vision. The example we have chosen will illustrate this.

This example has become one of the bases of the Third Industrial Revolution, as Jeremy Rifkin puts it. This is what he calls the Internet of Things, in its component where buildings can become the nodes of the energy and of the data networks of tomorrow. What is this concept based on?

In a Third Industrial Revolution approach, every building is a potential producer of energy it first uses for its own use (in self-consumption), and that can eventually be transmitted to others, through a smart grid. This exchange can be the subject of a commercial transaction between a seller and a buyer, based on a product (the energy supplied) whose cost can vary almost at any time. This last point thus implies implementing a dynamic system of data exchange which makes it possible to identify and establish a contractual relationship between producer and consumer, on the dual basis of demand and supply.

Referring to the current situation where self-consumption has entered the realm of possibilities for many buildings, where renewable energies are becoming more and more profitable and where buildings are gradually becoming more and more efficient in terms of energy efficiency, it is easy to imagine that reinjecting electrical energy into existing networks cannot continue, quite simply because these networks were not designed for this purpose. At the same time, the potential of smart grids, their capacity for resilience and flexibility represent major assets when it comes to the evolution of the energy transportation infrastructure, which would allow for the exchanges mentioned earlier. If we add the development of artificial intelligence in data management to the mix, we can highlight a global approach and the potential for efficiency and resilience that deserve to be assessed.

This concerns all buildings, regardless of their nature, especially when it comes to their ability to produce electrical energy (but we could later study the application to the recovery of wasted heat), their ability to generate savings and to increase the overall efficiency of a system through efficient management of all kinds of data (premises occupancy, production and consumption forecasts, contracting of exchanges, etc.). At the heart of all this is the end user, an individual who lives or works in the building under question, thus returning to the dual question of their behavior and their association with the initial reflection that relates to project co-development.

Implementing this concept today requires one to be equipped with the skills and technical means, but also being able to carry out the new developments, whether in the field of energy networks, data exchange or sociology. One needs a population which is preferably aware and ready to collaborate. Of course, one needs buildings on which energy production can be installed and connected with a new infrastructure, if the latter does not already exist.

Our thinking may be quickly directed to a city district, in which there are ongoing experiments taking place. But also to the academic world, which offers us the advantage of reaching a more immediate and straightforward mastery of the subject, because it benefits from being able to manage a global and unified group of stakeholders.

The major issues that we have identified are a wonderful field of action for our universities, those "micro-cities" which are largely confronted with the same problems as those that exist in everyday urban life. These problems directly involve young generations, but within a more territorially limited framework and with a different potential to take action. Could these institutions not now become the sites for fruitful reconstruction? What awaits us is an interesting exercise which is largely promising to test the transition model that has just been proposed.

Chapter 3 will allow us to appreciate how different universities around the world have become involved in global, innovative and responsible energy transition approaches. We will understand how they have, or have not, taken these "building blocks" into account, and we will be able to identify the common points, but also the differences in each approach, as well as the difficulties encountered over the course of a journey which is both demanding and rewarding.

3

University: The Ideal Place for Research and Implementation

3.1. Introduction

Today, there are hundreds of universities around the world that offer courses or specific approaches related to carbon neutrality, sustainable development and energy transition. But the number of those universities that have made a real commitment to carbon neutrality in their own operations is more limited.

There is a significant difference between those who share knowledge, arouse reflection and do research, and those who take the step of applying what they teach to their lives, by implementing practices on their own campus of research and creating clear demonstrations that relate to the energy transition. This may take the form of an environmental program with precise ambitions, commitments, working methods and governance.

It is not difficult to imagine the sheer richness of what could become an exemplary approach, but one that displays a coherence between words and actions, considering all that these approaches can bring to students, teachers and researchers, and more generally to the whole university and of course to the world that surrounds it.

This chapter presents the steps taken by six universities in Europe and North America to bring their campuses toward carbon neutrality, to raise awareness and involve users (students, staff, partners, visitors, etc.). In some cases, we see how these steps have transformed the campus into a full-scale living laboratory thanks to demonstrators who experiment with new technologies and new practices that contribute to the energy and societal transition.

3.2. Universities and transition: from university to univer'city

A university, with its students, faculty, staff, buildings, energy and water networks, and supply and mobility needs, is very often comparable to a small town. It is made up of several tens of thousands of people, sometimes with hundreds of buildings. These are places of work, residence, mobility and catering. But it is also a managed "territory" which benefits from a centralization of a certain number of data and functions, which simplify analysis and decision-making. A city does not have the same structures. To use the words of Pierre Giorgini, the university is also a city, and it is the univer'city.

It is in this respect that an analysis of the comments of universities engaged in transition approaches seems interesting to us. Of course, it is not our intention to offer an exhaustive review of all the universities that are investing in transition approaches here. But examining the plans of a few that have embarked on strong initiatives already proves to be very instructive. What areas do they want to focus their efforts on? What ambitions do they display? What resources have they allocated to their actions? What time horizon have they given themselves? What difficulties have been encountered and what are the limits to these approaches? And perhaps, what lessons can we apply to the city?

Because they display their ambitions clearly and in detail, we have selected the programs of the following universities:

– University of Manchester (UK);

– Stockholm University (Sweden);

– Boston University (USA);

– University of Reading (UK);

– University of British Columbia (Vancouver, Canada);

– Université Catholique de Lille (France).

The initial desire was not to focus mainly on the English-speaking world, but in the end these six universities were selected because they offered both a strong level of commitment and clear and accessible communication. Others, which are not presented here, also have programs of interest, such as University of Gloucestershire (UK), University of Edinburgh (UK), University of Leeds (UK), Nottingham Trent University (UK), University of California (USA), National University of Singapore.

We should also mention initiatives such as *Race to Zero Universities and Colleges*[1], which outlines the commitments made by 1,050 universities in October

1 Available at: https://www.educationracetozero.org/home.

2021 in Glasgow within the framework of the Times Higher Education Climate Impact Forum and UNEP [UNE 21], the International Universities Climate Alliance, which brings together 48 universities around the world[2]. Also worth mentioning is the launch of the World Alliance of Universities on Carbon Neutrality (WAUCN) on November 22, 2021, which in its 28 founding members includes 15 Chinese universities [WHI 21]. All of these initiatives converge toward an ambition to control the environmental footprint, and of course an international movement has been launched. This movement is expressed in multiple ways, covering commitments that differ according to the organizations, but which can be united under the common and unique banner of carbon neutrality. Individual objectives are specific, the goal can be understood and expressed in different ways, and the time horizon goes from 2030 (considering a reduced number of CO_2 emitting sources) for the most advanced, to 2050 for those who are less advanced, or who have a plan, but have not yet kicked off any concrete actions.

The next section of this chapter will briefly present the approaches of five universities organized around campuses which concentrate the majority of buildings and living spaces in a specific place within or on the outskirts of the city that hosts them. These universities aim to reduce their carbon emissions and move toward carbon neutrality, and seek to take ownership of the UN Sustainable Development Goals (section 1.3.4, Figure 1.9).

The case of the Université Catholique de Lille will then be dealt with in a specific section, due to the particularities of this university whose buildings are scattered in a district close to the center of the city of Lille and on the outskirts. In addition to aiming for carbon neutrality, it aims to transform the campus into a real-scale living laboratory because of demonstrators who experiment with new technologies and practices that should contribute to the energy and societal transition. The establishment of the university in the heart of the city and the fact that it is strongly integrated into it offer it the prospect of becoming a demonstrator of a sustainable and desirable district (small smart city).

3.3. Five universities moving toward carbon neutrality

3.3.1. *A reference framework for greenhouse gases emission sources*

Sources of greenhouse gas emissions are generally classified in a three-category reference framework called "scopes 1, 2 and 3". This international convention is used to calculate an organization's carbon emissions.

---

2 Available at: www.universitiesforclimate.org/.

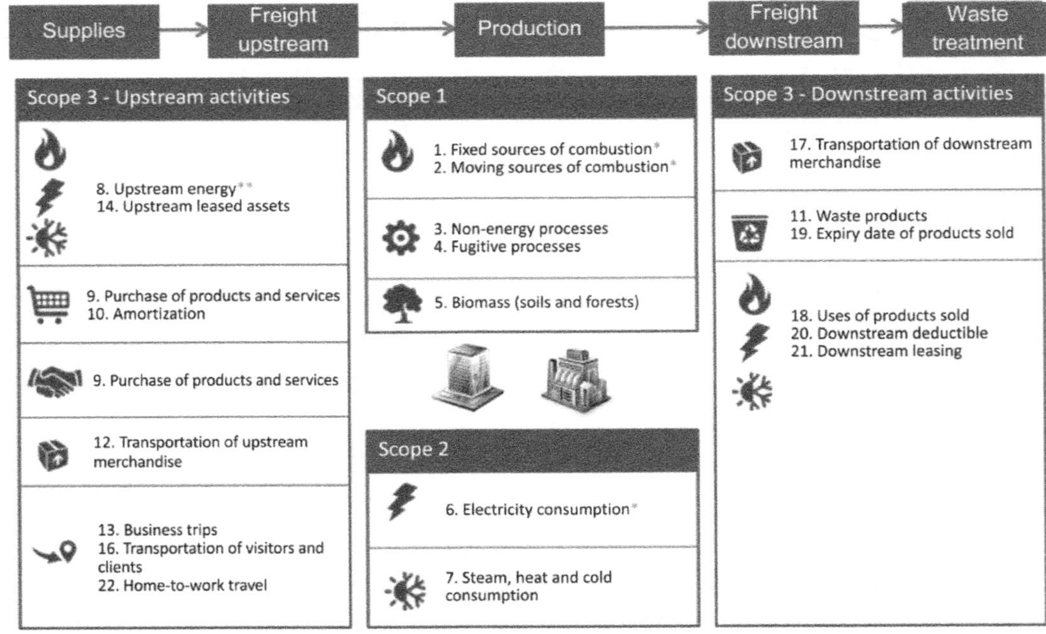

Figure 3.1. *Breakdown of an entity's emissions between direct emissions (emitted on site) and indirect emissions (emitted outside the site, but for its consumption) according to a diagram by Ademe (https://www.bilans-ges.ademe.en/static/images/basecarbone/fe_scope.png)*

Ademe proposes a definition of these different scopes as shown in Figure 3.1 [ADE 22]:

– Direct greenhouse gas (GHG) emissions (or scope 1): direct emissions from fixed or mobile installations located within the organizational premises, that is, emissions from sources owned or controlled by the organization, for example: combustion of fixed and mobile sources, non-combustion industrial processes, emissions from ruminants, biogas from technical landfills, leaks of refrigerants, nitrogen fertilization, biomass.

– Indirect emissions associated with energy (or scope 2): indirect emissions associated with the production of electricity, heat or steam imported for the organization's activities.

– Other indirect emissions (or scope 3): other emissions indirectly produced by the activities of the organization which are not counted in scope 2, but which are linked to the complete value chain, for example, purchase of raw materials, of services or other products, employee travel, upstream and downstream transport of goods, management of waste generated by the organization's activities, use and expiry of products and services sold, immobilization of property and equipment of production.

3.3.2. *University of Manchester*

Name: The University of Manchester

Website: www.sustainability.manchester.ac.uk

Number of students: 35,000

Teachers and staff: 11,000

Date announced for achieving carbon neutrality on scopes 1 and 2: 2038

Carbon emissions estimated at: 54,000 tons eq. CO_2 in 2018 (scopes 1 and 2)

Reference document: report on the environmental performance strategy, published in October 2020 [MAN 20]

Box 3.1. *Characteristics of The University of Manchester*

The carbon neutrality plan is one of the components within a more general environmental policy plan for sustainable development, which is in line with the 17

United Nations Sustainable Development Goals (Figure 1.9). This is obviously closely related to city policy (see section 2.3.2).

This sustainable development plan focuses on 12 major themes:

– management of carbon emissions;

– climate change awareness;

– construction and renovation of campus buildings;

– better use of water and energy resources;

– more “responsible” food;

– resource-efficient laboratories;

– a greener campus;

– a more responsible purchasing policy;

– research that provides solutions to environmental challenges;

– sustainable development throughout education;

– better control over business travel;

– reduction, reuse and recycling of waste.

The zero-carbon plan aims to develop a roadmap for the transition to a zero-carbon campus. It revolves around an objective of reducing direct (scope 1) and indirect (scope 2) emissions by 13% per year. This reduction is supported by a multifaceted plan:

– audits of existing buildings to assess renovation work, and a commitment that all new buildings are zero carbon;

– on-site energy production and the purchase of energy that is certified from renewable sources, up to 100% of demand;

– improvement of infrastructure;

– a commitment to put the zero-carbon approach at the heart of the programs and more generally in all the activities and policies of the university.

Scope 3 represents approximately 400,000 tons eq. CO_2, or 89% of total emissions, most of it coming from purchases. But the university recognizes the low availability of relevant data on this scope, so research is underway with external partners which aims to develop a better scientific approach to scope 3. The plan also defines measures to be taken to adapt to climate change.

The university is committed to review its investment strategies in line with the carbon strategy and to work with decarbonization partners in Greater Manchester.

Figure 3.2 shows the trajectory of direct emissions as desired (in red), compared to reality (in blue) and the consequences of any deviations from the objective.

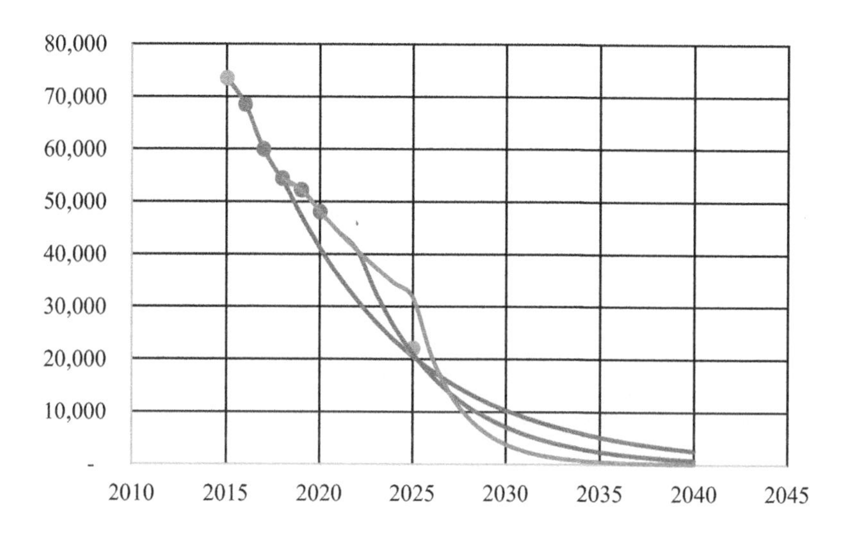

Figure 3.2. *University of Manchester: possible trajectories of carbon emissions (scopes 1 and 2). Blue curve (until 2020): achieved. Red curve: initial target of 13% reduction per year. Purple curve: if corrected – 20% from 2022. Light blue curve: if corrected – 35% from 2025 [MAN 20]. For a color version of this figure, see www.iste.co.uk/robyns/smartusers.zip*

In summary, important work has been launched, and it is well integrated into a very broad global approach, but this work sometimes comes up against a lack of data. It demonstrates a desire to tackle all the positions corresponding to scopes 1 and 2 and to find solutions to assess and control scope 3, most likely with a little time lag.

3.3.3. *Stockholm University*

The university has defined a roadmap for the 2020–2040 period. In his introductory remarks, the author clearly expresses the need for the university to reduce its GHG emissions to guarantee its own credibility. He places this approach within a national perspective, but also within that of the European Green Pact (*Green Deal*).

Name: Stockholm University

Website: www.su.se

Number of students: 29,300

Teachers and staff: 7,100

Date announced for achieving carbon neutrality: 2040

Carbon emissions: not communicated

Reference document: *Stockholm University Climate Roadmap for the Period 2020-2040* [STO 22]

Box 3.2. *Characteristics of Stockholm University*

The university has implemented environmental management in accordance with the ISO 14001 standard, and bases its reflection on the 17 Sustainable Development Objectives defined by the UN (Figure 1.9).

Carbon neutrality is approached from the double angle of positive emissions and negative emissions, with intentional work carried out on these two points, and of course with the objective that the sum of the two will be zero. The first observation is that the data available are insufficient and that a broader review will be necessary, with the possibility to discover new sources of emissions not listed up to that point.

In Stockholm, scopes 2 and 3 represent approximately 95% of total emissions, and it is on the corresponding items that our efforts will be focused. The main measures listed and published to date are as follows:

– Scope 1: replacement of internal combustion engine vehicles with electric or gas vehicles. Analysis of leaks from cooling systems.

– Scope 2: continuous increase in energy efficiency, feasibility study of an urban heating network, incentive for investment in solar energy by building owners.

– Scope 3: this is where the main effort will be made:

- business travel: define a new policy, take advantage of the lessons learned from the pandemic, consider using the train with travel agencies;

- purchases of goods and services: impact assessment, requiring external partners to turn to low-emission transport, evaluating user behavior and creating a circular economy offer for furniture and material goods;

- logistics: development of a low-carbon policy in the purchase of goods and services;

- assets under management and investments: review of invested capital, continue to separate from assets linked to fossil fuels;

- rented buildings and new or renovated buildings: assess the possibility of calculating emissions in the design phases of new or renovated buildings, define the share due to the owner and the share of the tenant, seek possible improvements related to efficiency, include "green" clauses in leases being renegotiated;

- waste management: reducing quantities, recycling furniture;

- external services: facilitate soft mobility for students and include requirements for catering and partner businesses on campus.

The university observes that from the moment scope 3 is taken into consideration, some GHG emissions cannot be avoided by 2040, such as those linked to the use of aircraft, hence the need to include carbon mitigation and offsetting processes (negative emissions) in the plan. This subject is recognized as still being imprecise, so it must be regarded in an exploratory sense. The possibility to increase CO_2 capture through green spaces, to produce bioenergies locally, to contribute to the development of greater efficiency in the use of resources and to reduce GHG emissions is mentioned by the society. The negative emissions approach is in the research stage, and it has not yet been materialized into action.

At the same time, the university places its contribution to a more sustainable and more efficient society in the management of resources which are at the forefront as part of its societal action plan, through a very comprehensive program of teaching and research. It states that research studies will be "highly probable" and that the educational work carried out on a daily basis will have a positive influence on a more sustainable development of society, in particular with regard to the 17 objectives of sustainable development.

Finally, it should be noted that an "environmental council" is in charge of monitoring actions with milestone checks every 2 years. The first priority is to review current broadcasts. Covering 20 years, the climate plan will be piloted and will display many ambitions. The commitment is clear, but there are still many questions to be resolved.

3.3.4. *Boston University*

This university established its climate plan in 2017, and among the five universities selected in this work, it is the one that most insists on the need to adapt to climate change, in particular because the location of some of its buildings makes it very sensitive to flooding. This action plan precisely identifies the costs and benefits of the transition to carbon neutrality and underlines the contribution that this approach will have in terms of research and education. The whole presentation that is given about the project begins with "act now".

Name: Boston University

Website: www.bu.edu

Number of students: 33,500

Teachers and staff: more than 10,000

Date announced for achieving carbon neutrality on scopes 1 and 2: 2040

Carbon emissions estimated at: 129,400 tons eq. CO_2 in 2016 (scopes 1 and 2), an order of magnitude of 200,000 t eq. CO_2 is given for scope 3

Reference document: *Recommendations of the Climate Action Task Force for Boston University's Climate Action Plan* [JAN 17]

Box 3.3. *Characteristics of Boston University*

An "inventory" shows that the use of electricity, natural gas and steam represents the majority of scopes 1 and 2 emissions. The actions that have been underway since 2016 would lead to a 20% reduction in the 2050 horizon. Scope 3 can be approached in several ways, for lack of sufficient data, and is estimated globally between 178,000 and 226,000 tons eq. CO_2 for a final retained value of 202,000 tons eq. CO_2. Within this total, the impact of displacements is assessed for 30,800 t eq. CO_2, to which emissions related to all purchases and waste have been added.

The working group (task force) studied three scenarios: *BU good, BU better* and *BU bold*. Each scenario focuses on the same axes: providing electricity from renewable sources, investing in energy efficiency for heating, ventilation, air conditioning and lighting. The only difference between the three scenarios is the level of ambition, which increases from good to bold. The bold scenario is retained by the working group. Figure 3.3 summarizes the strategy proposed in this scenario for the decarbonization of energy sources.

The sharp drop in electricity-related emissions corresponds to the decision to only buy electricity from renewable energy suppliers, and to do so immediately, which means that they could benefit from federal subsidies. A rather complex certification system has been put in place, which should make it possible to accurately track the origin of the energy purchased.

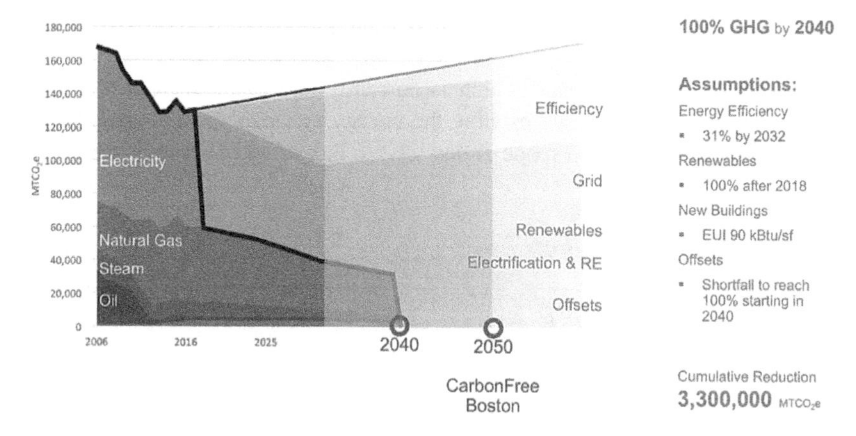

Figure 3.3. *Bold energy scenario for Boston University [JAN 17]. For a color version of this figure, see www.iste.co.uk/robyns/smartusers.zip*

While work to improve the energy efficiency of existing buildings is mentioned (via the orange area of the graph), there is no mention of local energy production, except perhaps for new buildings. The dark green area in Figure 3.3 corresponds to the transition to electricity from buildings heated by fossil fuels, the lighter green area is the purchase share mentioned above and the pale green area illustrates the decarbonization of electricity planned in the New England region. Finally, the pink area corresponds to the purchase of carbon credits to offset the balance. This strategy makes targeting carbon neutrality by 2040 for scopes 1 and 2 possible.

The first observation that can be made on scope 3 is the lack of reliable data, and there is little monitoring of the quality of emissions within this scope. The major components can nevertheless be identified, and the working group makes the following recommendations:

– Mobility: to carry out additional studies on the means of reducing corresponding emissions, which are considered intrinsically necessary for the university to function. This approach is complemented by improved urban travel and the purchase of electric vehicles for the campus. The university emphasizes the need to work closely with the city of Boston on these topics.

– Purchases and waste streams: here too, the university comes up against a lack of reliable data; however, a zero-waste certification is proposed, but the first effort will be to set up a data collection process.

As mentioned, the climate plan is supplemented by a comprehensive approach based on the resilience of the campus in the face of two natural phenomena, the probability of which increases over time: flooding of some of the buildings located near the *Charles River*, itself leading to the port, and urban heatwaves.

All in all, a structured plan, which is currently being put in place for a 2040 objective, is significantly based on how the energy mix evolves and asks the same questions as other projects on scope 3: how is it to be evaluated and put into action?

3.3.5. *University of Reading*

Name: University of Reading

Website: www.reading.ac.uk

Number of students: 22,480

Teachers and staff: not specified

Date announced for achieving carbon neutrality: 2036

Estimated carbon emissions: 43,984 tons eq. CO_2 (2008–2009)

Reference documents: *University of Reading Net Zero Carbon Plan 2021–2030* [FER 21]

Box 3.4. *Characteristics of the University of Reading*

Unlike the three previous universities, Reading started reducing its carbon emissions from 2009 onwards, and the reduction rate seen in 2021 is 44%. The objectives, which are reviewed every 3 years, are currently included in a 2020–2026 strategic plan. Carbon neutrality is described as a balance between an equilibrium of positive and negative emissions. The university affirms that it will take all indirect emissions of scope 3 for 66 to 100% of their contribution into account. The route it envisions to the year 2036 is summarized in Figure 3.4.

In 2020, carbon reduction objectives relate, according to the scopes concerned, to the following subjects:

– scope 1: fossil fuels used, university vehicles, refrigerant gases;

– scope 2: electrical energy used in buildings;

– scope 3: business travel, water and waste, on-campus halls of residence.

What will be considered in a second step (estimate of impacts and actions to be taken) is as follows: scope 3, waste from construction, purchasing policy, off-campus residences.

Although qualified as "out of scope", the topics that will be considered later are student travel, agricultural land and livestock.

The university has a very clear view of what still needs to be done, which can be summed up in six main points, while not avoiding the fact that the path to come will not necessarily be easy to follow:

– Replacing all heating systems. As many buildings are still heated with fossil fuels, various possibilities are being considered, in particular the renovation of existing heat pumps as well as more effectively monitoring the installations. This is the most important part of decarbonization.

– A more carbon-free electrical energy supply. The university buys part of its electricity which is "guaranteed to come from renewable sources", but it doubts the true effectiveness of this approach and therefore plans to supplement it with direct purchases from a renewable energy producer.

– An increase in on-site renewable energy production is planned with the corresponding installation of storage capacities.

– An improvement in energy efficiency which will concern lighting, ventilation, air conditioning and IT.

– In terms of travel, the objective is to cap emissions related to business travel by reconsidering the use of plane journeys whenever possible.

– In light of the fact that there will be carbon emissions that cannot be eliminated, the program accounts for the purchase of carbon credits and the development of natural carbon sinks on the university's land holdings. This is also seen as research work.

Here, we can find themes of actions already seen elsewhere, in a plan that is already being implemented and therefore inevitably relates to very specific objectives. The plan mentions financial results with savings of 34 million pounds sterling, or approximately 40 million euros, and future investments of around 60 million euros.

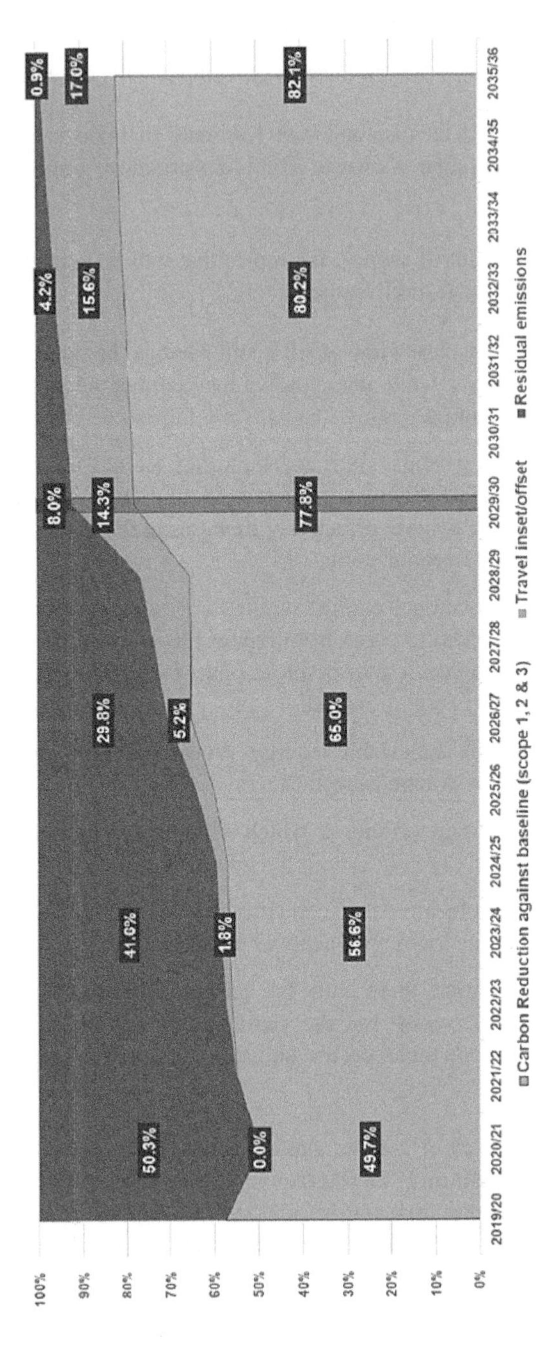

Figure 3.4. *Evolution of planned carbon emissions reduction (green: direct actions, gray: capture and offset, blue: residual) from the University of Reading [FER 21]. For a color version of this figure, see www.iste.co.uk/robyns/smartusers.zip*

3.3.6. *The University of British Colombia*

Name: The University of British Columbia

Website: www.ubc.ca

Number of students: 68,500

Teachers and staff: 17,800

Date announced for achieving carbon neutrality: 2030–2035

Carbon emissions estimated at: 46,553 tons eq. CO_2 in 2018 (scopes 1 and 2)

Reference documents: *UBC Vancouver Climate Action Plan 2030* [UNI 22]

Box 3.5. *Characteristics of The University of British Colombia*

The first climate plan of this university dates back to 2010, and all the documents related to it relate to a genuine and deep awareness of climate change, its origins and its consequences. A second plan was made in 2020. Between 2007 and 2018, the reduction in carbon emissions on the Vancouver campus was 38%, while at the same time the floor areas (+21%) and the number of students (+32%) increased.

The 2030 plan places the university in a process of acceleration as it relates at the same time to buildings, energy supply and other impacts in terms of GHG emissions from activities.

This plan covers seven main principles:

– Operation of the campus (100% low carbon objective by 2030): this concerns the energy used by the buildings with the suppression of fossil fuels, a decarbonization plan that relates to the maintenance and renovation of the buildings, the evolution of a hot water network, etc.

– Travel (target –45%): development of teleworking and carpooling, “pass” for students while waiting for a dedicated public transport in 2030. Development of charging stations, encouraging the use of bicycles.

– Food (target –50%): development of a global approach at campus level, purchasing policy, waste management and creation of a Climate Friendly Food label. The importance of this item in the overall carbon balance (scopes 1, 2 and 3) is clearly highlighted in Figure 3.5. Also note the importance of scope 3, which represents approximately 70% of total emissions.

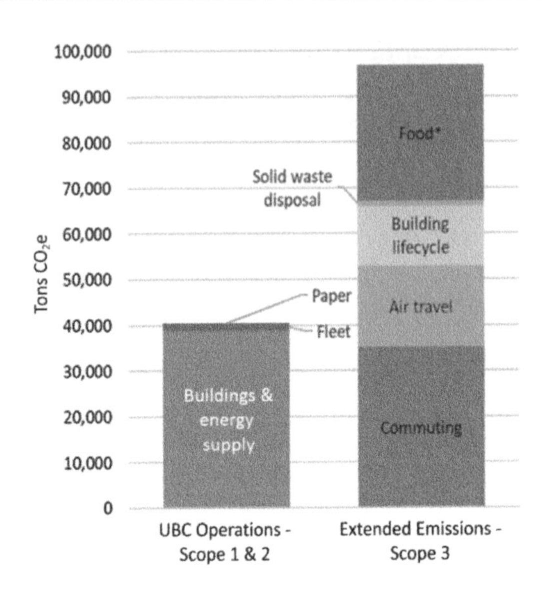

Figure 3.5. *GHG emissions in 2019 by origin from The University of British Columbia [UNI 22]. For a color version of this figure, see www.iste.co.uk/robyns/smartusers.zip*

– Air transportation (objective –50% low carbon in 2030): reduced travel related to congresses and conferences or use of other means of participation. The plan focuses on behavior change, but requires additional studies.

– Waste and materials (50% reduction in waste): an update of a zero-waste plan, a consideration of the circular economy and identification of obstacles.

– Carbon footprint of materials (target –50% in 2030): this involves controlling the carbon footprint of construction materials used for new buildings, but also renovation. This objective is made up partly of research and partly of application and limitation in the use of certain materials.

– Individual commitment (objective: participation of 75% of teachers, staff and students): train, inform and communicate, including externally. To do this, develop methods and resources.

Aside from this plan, there are also plans that have been defined for residences linked to the campus, an action plan for buildings in general, a plan to manage water linked to storms, a plan to preserve water resources, as well as for adaptation and resilience plans, etc. The set of actions is very complete and the organization required for these multiple objectives is in place.

Here, we are looking at a different scale compared to other universities. Similar to Reading, this plan covers the majority, if not all, of the areas which allow us to achieve carbon neutrality for the university. We can recall that Vancouver is a city that has long had a strong ecological commitment, and it is therefore not surprising to find that the university is one of its essential driving forces. The fact remains that this plan is remarkable, and it demonstrates the results that a committed path to carbon neutrality can bring.

3.3.7. *Summary*

These five universities are committed to a carbon neutral approach, with five different proposals, and a lot of lessons to be learnt.

The common thread, or the reference, is of course scopes 1, 2 and 3, along with the understanding and interpretations that we can make of them. All of the universities emphasize the difficulty of having reliable data, or even simply available data, in particular when it comes to scope 3, which represents between 60 and 90% of emissions. The universities that have started first, those which have already made good progress in emission reductions linked to scopes 1 and 2, have the clearest vision of what is possible to achieve in scope 3.

The notion of carbon neutrality is at once completely accepted, even brought to light in the communications from universities, yet it also remains a difficult objective to achieve. The proposed measures are different each time because they are closely linked to the characteristics of each location. Stockholm talks little about scope 1 because this subject has largely been resolved at the national level. Rather, the origin of electricity, which is often very carbon intensive at the start, is a major subject. Many but not all talk about a renewable energy development program on their site. The Anglo-Saxon mode of operation involves purchasing energy from independent producers.

Setting a precise, quantified objective in terms of carbon neutrality constitutes a powerful driving force when accompanied by attentive governance. These universities have a structured plan with strong governance, and have shown, at least those that have already made good progress, the ability to ensure follow-up and corrective measures where necessary. It is never a "long calm river", but a course that requires permanent monitoring and the capacity to react when needed.

Scope 3 is always highlighted as difficult to assess, and it is also the one which is most difficult to enact. There are items that are common to all universities: transport, purchases, food and waste. Universities are developing compensation programs, purchasing carbon credits or creating carbon sinks on their land holdings to deal with what cannot be removed, and we are aware of the random nature of this notion of compensation.

Among all of this, there is a constant: it takes 20 years, sometimes a little more, to achieve carbon neutrality. This is both long when we place this duration in the context of the climate urgency, and short when the we consider everything that needs to be done, particularly when it comes to the collection of data, and everything that needs to be reformed over the course of the operation. This duration can be compared to that which has been identified for cities in Chapter 2, being of the same order.

In the most advanced plans, particularly in Vancouver, we can see in the documents that climate action has been integrated at the heart of the mission that the university has given itself, to the point of becoming its primary subject. Vancouver has publicly announced that it is signing the climate emergency declaration at the end of 2019 along with 1,700 other institutions around the world. Climate action is becoming a major priority for the university in a structure of teaching, research and development, exemplarity, of participation in society, progress and leadership. It impacts all departments, without exception. Finally, it is an extraordinary way of promoting the university to students and civil society alike.

3.4. The Live TREE program from the Université Catholique de Lille

3.4.1. *The specificities of the university*

All companies, associations, organizations, etc., must change their practices in order to preserve the planet and its ecosystems, and to achieve carbon neutrality. The university, and more generally all training and research establishments, must go further by raising the awareness of young people in particular, by enabling them to build the futures that they deem desirable for their children, by disseminating good practices and leading by example.

In November 2013, the Université Catholique de Lille[3] decided to commit itself resolutely to the Third Industrial Revolution theorized by the American economist Jeremy Rifkin [RIF 12] (see section 1.3.3), and thus to contribute to the essential energy, ecological, economic and social transition of the university and the Hauts-

3 The Université Catholique de Lille is a multidisciplinary federation that brings together faculties, colleges, engineering and business schools, and research centers on the one hand, and a hospital group on the other. At the start of the 2021 academic year, the Université Catholique de Lille had welcomed 38,500 students in more than 300 training courses, and 7,000 staff members.

de-France region by developing an orientation plan for the “Live TREE” program (Lille Vauban-Esquermes in energy, ecological and economic transition).

The Live TREE program aims to set up a global approach which concerns all aspects of university life, and is connected to its environment: students, staff, inhabitants of the Vauban district, partner companies, communities, associations, etc. Through Live TREE, the university wishes to strengthen its role of providing “global education” for students as well as its social responsibility as a university; to engage its students, once they have been launched in their adult and professional life, to take action and build new models of a society that is more respectful of the environment and of the individual human, one that is more just and equitable.

Live TREE is therefore a global project, which crosses the various functions of the university and which implies a transformation of the modes of governance and management used by the university.

Pierre Giorgini, the president of the university between 2012 and 2020, offers [GIO 16b] his vision for the university of the future which reinvents itself: “Everywhere, universities are becoming genuine laboratories to ‘recreate the world’. They are ‘zero carbon’ and offer students, whatever their level and age, a stable creative environment in which they can experiment with new techniques, new social and economic postures on campus, and eventually invent the new world. Social shops have been set up, where researchers in the humanities, students, economic and societal stakeholders coexist. They become spaces for creation, experimentation and scientific observation of the initiatives which are analyzed and leveraged in transdisciplinary research-action approaches (experimental economics, for example). Workshops and fablabs are open not only to students, but also to anyone, regardless of their level of training; who has an idea, a talent, a desire to learn; to manufacture; or to create an activity (company, action, community). They are equipped with 3D printers, laser cutters, wood lathes and the most sophisticated machines [...]. Creation incubators (companies, associations, start-ups, entrepreneurship) are installed nearby and brought to life. The university becomes a veritable hive, an agora of chance encounters.”

The buildings of the Université Catholique de Lille have been built from 1877 until today following the creation of new establishments and their expansion over time. The architecture of these buildings is therefore very variable depending on the period of construction concerned. These buildings are largely scattered throughout the city in the Vauban district of Lille, with certain high concentrations in the form of islets, and in other districts of the metropolis such as Roubaix (EDHEC business school, Pôle IIID graphic animation school), Ennetière-en-Weppes, Lomme (Saint-Philibert hospital and Humanicité district) and other districts of Lille such as Porte des Postes (establishment accommodation for dependent elderly people, EHPAD Féron-Vrau) and Porte de Valen-ciennes (Saint-Vincent hospital).

The university campus therefore has a specific configuration which is due to the dispersion of buildings within a district, meaning that the buildings and blocks of buildings are linked together by public energy networks (electricity, heat, gas). This is contrary to a more traditional campus where all the buildings are concentrated on the same geographical entity and therefore linked together by private energy networks. It should be noted that within certain blocks, the university owns private electricity networks.

Based on these objectives and the geographical context of the university, the Live TREE program is organized around several themes:

– carbon neutrality of the university;

– living lab: experimental buildings and living spaces in a sustainable and desirable district;

– Vauban-Esquermes zen campus:

  - openness to the neighborhood and third places,

  - green mobility,

  - climate and nature in the city;

– student experience: student training and involvement, learning campus;

– transdisciplinary research between human and social sciences and engineering sciences;

– communication and international outreach;

– dissemination and foresight.

These themes make it possible to meet a large number of the UN's sustainable development goals (section 1.3.4, Figure 1.9).

3.4.2. *The meaning of a program*

Beyond the technical transformation of the university's buildings, the origin of the energy it consumes and the means of transportation it uses, the success of the energy and societal transition can only be achieved if all actors are made aware, feel concerned, see themselves as actors, find meaning and start adapting their actions. In this way, the actors of the Live TREE program regularly question the meaning of their actions. Because we know that it is a question of "promoting awareness of a common origin, of mutual belonging and of a future shared by all" [FRA 15], the university must go further. Luc Dubrulle explains the following: "Being an interdependent human is a fact we must always work harder to understand, but being truly human is a duty. It implies work of a scientific nature to better understand ourselves, as well as ethics which guide us to act in a better way. This work must always be revisited, but the initial periods are particularly informative because they forge men and women who will decide to act throughout their lives with the intention of taking care of our common home, to varying degrees. The university's ambition is therefore to enable its 38,500 students to graduate with ecological virtues that will impel them to act throughout their lives, both in their personal and civic lives and especially in their professions which they will occupy" [DUB 17].

As has already been underlined, interdisciplinarity is also essential for a successful energy and societal transition. The university therefore has an important role to play in promoting it: "A dialogue between the sciences themselves is also necessary because each is used to confining itself within the limits of its own language, and specialization tends to imply isolation and absolutization of each person's knowledge. This prevents us from properly addressing environmental problems" [FRA 15].

To succeed in its contribution to the global energy and societal transition, the university must therefore:

– ensure that future generations, that is, the 38,500 students of the university are at the heart of this transformation, by setting up specific activities that encourage student initiatives, and by supporting these initiatives;

– bring the different disciplines of the university together, in order to develop interdisciplinary approaches linking human and social sciences and engineering sciences, essential for a successful transition everybody feels a sense of ownership of, without forgetting the most vulnerable individuals.

Gandhi's famous quote is very relevant here: "Let us be the change we want to see in the world".

3.4.3. *Carbon neutrality: a difficult equation to solve*

Carbon neutrality assumes that the CO_2 emitted is practically equivalent to the CO_2 absorbed; the best thing is not to emit CO_2 in the first place! Figure 3.6 represents an estimate of the distribution of the university's CO_2 emitting stations in 2019 in the Vauban district, that is, approximately 20,000 students and 4,500 staff members, for a total of between 30,000 and 35,000 tons of CO_2 (excluding the hospital, catering and housing sectors). The most significant items are energy consumed, representing 14%, and travel (students, staff, etc.), generating 61% of emissions. We also note the contribution of digital technology via the computer hub and the associated electrical equipment, which stands at 3%.

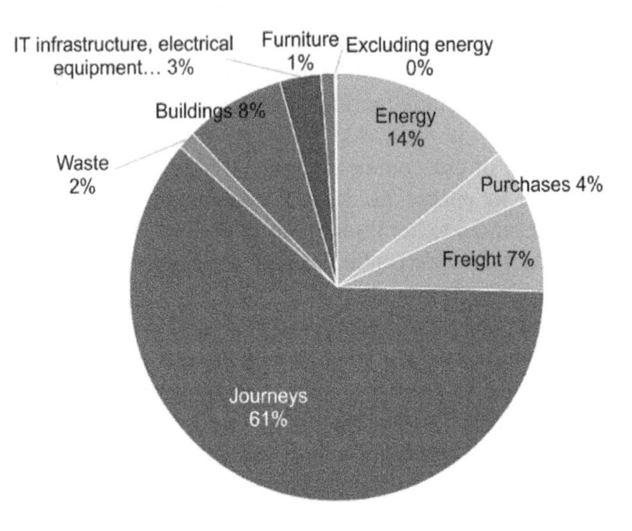

Figure 3.6. *Estimation of the distribution of CO_2 emitting stations from the Université Catholique de Lille. For a color version of this figure, see www.iste.co.uk/robyns/smartusers.zip*

The Japanese economist Yoichi Kaya's equation which was presented in Chapter 1 (section 1.2.4) could be applied to the university context:

$$CO_2 = Population \times \frac{Products}{Population} \times \frac{Energy}{Products} \times \frac{CO_2}{Energy}. \quad [3.1]$$

The first term corresponds to the university population: students, researchers, teachers, caregivers, administrative and technical staff, etc. The majority of universities seek to develop specifically by increasing their number of students and researchers.

The second term represents the products of the university divided by its population. Any university seeks to increase its income through student registration fees, research projects, the promotion of its activities, etc., and therefore wants this ratio to grow, that is to say, to higher than 1. Increasing activity generally leads to an increase in CO_2 emissions (more space occupied, travel, consumption of materials and energy, etc.).

The third term represents the university's energy efficiency, but also its level of sobriety after adjusting behavior. We imagine that this efficiency level can be improved and that this term can decrease below 1 like the global ratio (illustrated in Figure 1.5). However, we must monitor two consumption items which tend to increase and for which a certain level of sobriety is to be defined: international travel, which has a logical tendency to increase with internationalization training and research which is encouraged, and the digitization of many activities, including training. The development of digital technology makes it possible to reduce travel and can therefore contribute to a certain sobriety, but it is also energy-intensive (section 1.4.5), which means a compromise must be sought. As a reminder, travel already accounts for two-thirds of the university's CO_2 emissions, as shown in Figure 3.6.

The fourth term represents greenhouse gas emissions, mainly the CO_2 emitted by the primary energy sources consumed. In order to reduce it, it is necessary to exploit renewable energy sources and/or to guarantee an electricity supply, heat and low-carbon gases. The wish to act on this term encourages the self-production and self-consumption of renewable energy, as well as the organization of local energy communities. Due to the location of the university, being in the heart of the city, photovoltaic production is deployed as much as possible on the roofs of buildings.

Given that the university would not like their performance to decline, the action levers will have to reduce the two terms on the right of the equation which concern efficiency and energy sobriety as well as carbon-free energies. If acting on these two terms is not enough, it will be necessary to offset CO_2 emissions by investing in carbon sinks, that is, essentially by reforesting land. However, the concept of carbon neutrality is limited when applied to a company or a university because this entity works in synergy with other actors in an ecosystem or a territory. Ideally, carbon neutrality should be sought at this scale; entities do not have full control over public transport solutions that are offered, the energy sources that are used for heating and electricity networks, etc.

An initial target to reduce carbon emission was applied to the historic quarter (Figure 3.7) which combines historic and partly renovated buildings such as an academic hotel dating from 1877, the Rizomm building belonging to the Faculty of Management, Economics and Sciences and the Junia School of Engineering. The first carbon balance had to be carried out. This work was carried out with Dalkia's support, taking Ademe's scopes 1 and 2 into account (section 3.3.1, Figure 3.1), in other words, the energy consumed. The studies carried out have shown that reducing carbon emissions linked to heat will be achieved by renovating buildings, reducing carbon emissions from the Résonor heating network managed by Dalkia by two-thirds, and by raising awareness and involvement among users, with a target energy gain of 10%. The same is true for electricity-related emissions, which will be reduced by renovating buildings so that they become smart, by producing self-consumed photovoltaic energy locally, and by raising awareness and involving users, with a target gain of 10% once again. The quantities of CO_2 emitted in tons are illustrated in Figure 3.8, which highlights the fact that 30% of the emissions still need to be offset by planting trees, but also by renovating or rebuilding other old university buildings by making them energy positive, in other words by making them generate more energy (of a renewable origin) than they consume over the course of a year, offsetting the emissions of other buildings or blocks of buildings when carbon neutrality proves difficult to achieve.

Figure 3.7. *Image of the historic quarter of the Université Catholique de Lille (source: UCL Media Lab)*

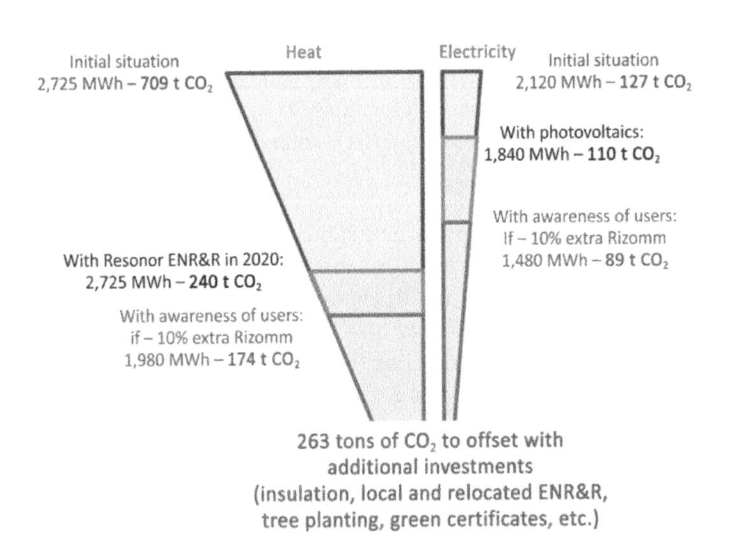

Figure 3.8. *Tons of CO_2 emitted by energy consumption in the historic quarter of the university and the strategy to reduce emissions (ENR&R = renewable and recovered energy)*

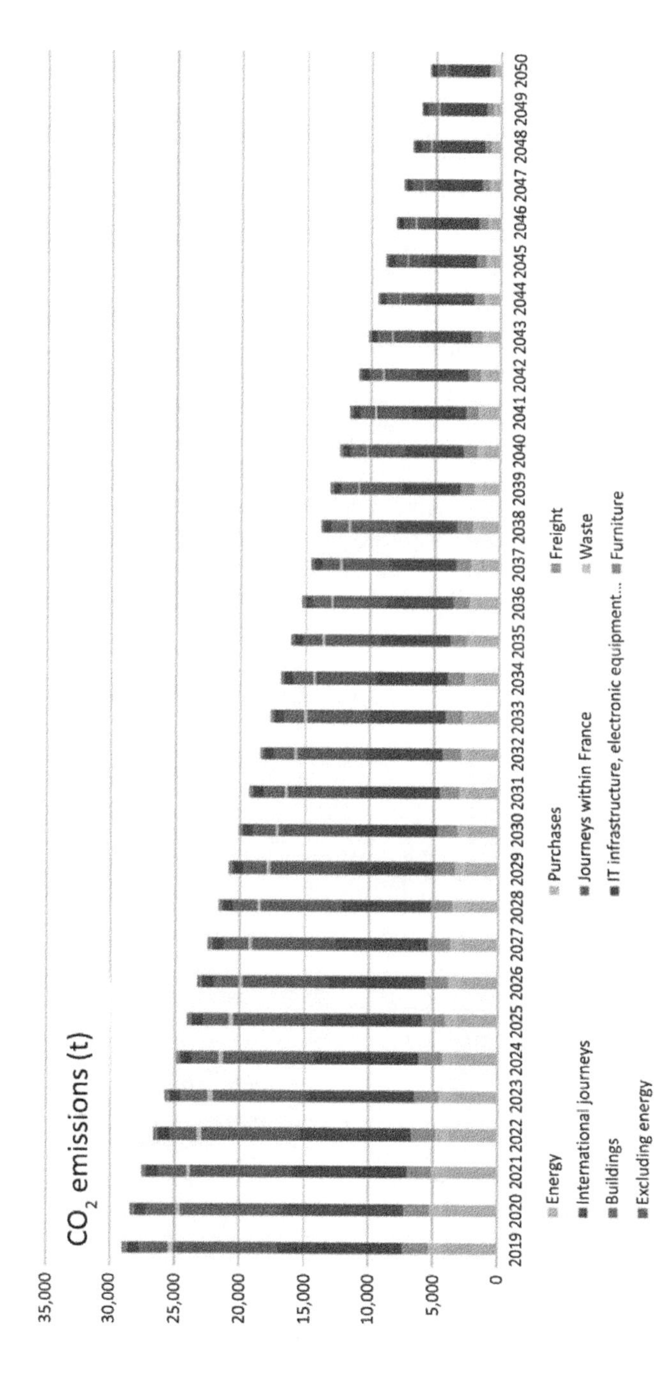

Figure 3.9. *Ideal trajectory for reducing the university's carbon emissions by 2050 (source: Grégoire Destombes). For a color version of this figure, see www.iste.co.uk/robyns/smartusers.zip*

Figure 3.9 depicts an ideal trajectory to reduce the university's carbon emissions by 2050, to contribute to the goal of limiting the increase in global temperature to between 1.5 and 2°C [INT 18]. The objective is to get as close as possible to carbon neutrality by 2050 while respecting a carbon budget that is consistent with the remaining carbon budget given by the IPCC (Intergovernmental Panel on Climate Change). This carbon budget is represented by the area shown by the trajectory in Figure 3.9.

The construction of this trajectory is carried out with all the entities of the university. It requires the sharing of experience, as well as the development of research projects to find solutions to specific problems, for example, in the hospital sector.

But the achievement of this trajectory also presupposes that all university actors can mobilize themselves (students, staff, partners, etc.) and work with local authorities. Indeed, the university's carbon footprint will be the result of the contributions and interactions between the three groups represented in Figure 3.10: the university establishments, the actors (students, staff, partners) and the local authorities.

At the territorial level, the city of Lille has drawn up a low-carbon pact signed by around a hundred companies and organizations located around the city, including the Université Catholique de Lille. The European Metropolis of Lille has drawn up a territorial climate-air-energy plan (PCAET) in which the university is a stakeholder.

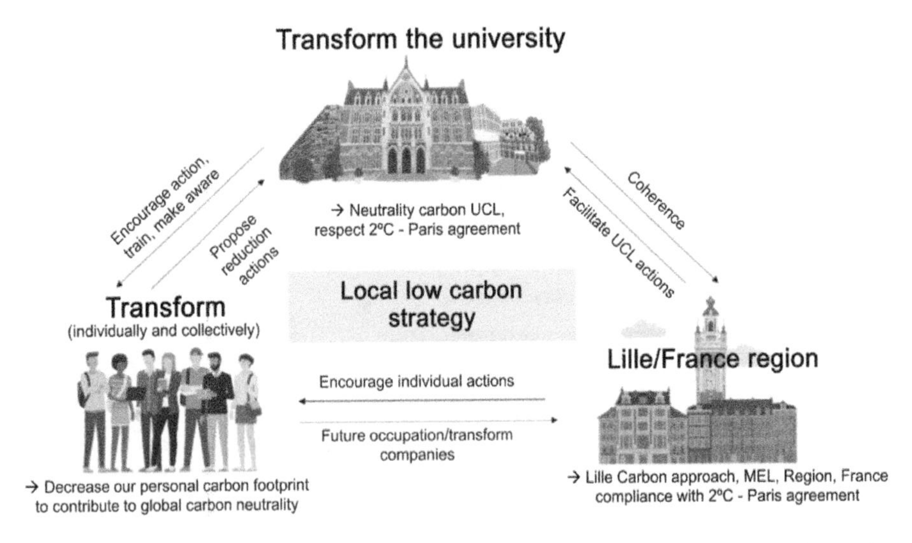

Figure 3.10. *The carbon footprint of the university will be the result of the contributions and interactions between the three groups: university establishments; student actors, staff, and partners; and local authorities (source: Grégoire Destombes)*

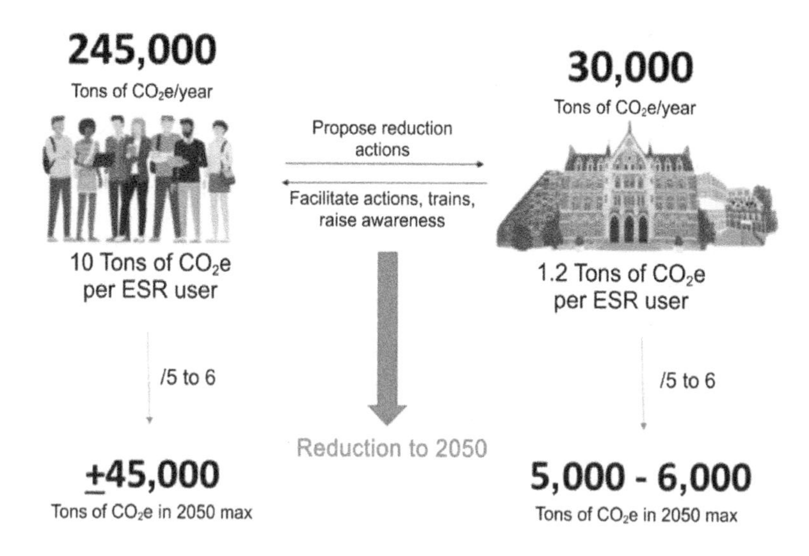

Figure 3.11. *Interactions aimed at informing, raising awareness and mobilizing university stakeholders to divide their carbon footprint by 5 to 6 by 2050 (ESR = higher education and research) (source: Grégoire Destombes)*

Through its mission of training and education, the university has a special responsibility and opportunity to raise awareness among its students and staff so that they contribute to reducing the university's carbon footprint, but also to reduce their personal carbon footprint in their daily activities in the process. Figure 3.11 illustrates this interaction aimed at informing, raising awareness and mobilizing university players so that they divide their carbon footprint by 5 to 6, which amounts to going from 10 tons of CO_2/year in 2020 to approximately 2 tons of CO_2/year in 2050. An important step in this mobilization is the organization in the fall of 2022 of a University Convention for the climate, the aim of which is to allow students and staff to propose actions that contribute to the development of the university's carbon trajectory which aims for neutrality in 2050, involving actions and visions that transform habits and behaviors, as well as ways of seeing and being in the world.

3.4.4. *Demonstrators*

3.4.4.1. *Necessary demonstrators*

As explained in the first chapter (section 1.4.9), implementing the energy and societal transition will involve experimenting with numerous technological and economic solutions on a real scale, with local societal organizations, etc. It will be important to accept trial, error and adjustment and to assess how viable and effective the technologies are, as well as their impact on the environment and their social acceptance, which is not a given. One must find new economic models to design new legal frameworks, etc. It is also about convincing people in order to limit resistance to change. Thus, it is a question of leading by example, which is exemplary.

Among the objectives held by the Live TREE program, the aim is to test tertiary and residential buildings in real conditions to the extent that they might become energy-smart by integrating socio-technical dimensions with user involvement. These buildings interact with energy networks that are themselves becoming intelligent, integrating new practices of energy self-production and self-consumption. All of these systems are interconnected by information systems that induce a digital energy convergence between smart buildings, smart grids, the Internet of things and people. At the scale of the campus and therefore of a district, an urban experiment has been initiated: the premise of a potential “smart city” demonstrator. The Live TREE program therefore incorporates a strong digital component into its approach to explore the transition. Thus, the university creates an Internet of Everything by connecting its buildings and users while bringing their data together in a dedicated and entirely local informational system. The purpose of this system is to allow real-time management of the mini-city that is the campus. It also manages the electric smart grid which is made up of the various solar energy production plants. It is about prefiguring all needs of the city of the future by highlighting all of the problems that the population will inevitably encounter in smart city type projects.

3.4.4.2. *Smart buildings*

In France, in 2016, residential and tertiary buildings accounted for 45% of the total energy consumed. The French PACTE law of 2019 aims to reduce the consumption of buildings by 40% by 2030 and by 60% by 2050.

Buildings will therefore have to considerably improve their energy performance. They will also play a fundamental role in the development of smart energy networks, energy micro-grids, eco-districts and smart cities. Figure 3.12 illustrates different characteristics of a smart building, which integrates local energy production and storage, controllable stations whose consumption levels can be modulated (lighting, heating, electric vehicles, etc.), which can be connected to the electrical distribution network as well as to external sources or can operate in isolation, that is to say, disconnected from the electrical grid [ROB 19].

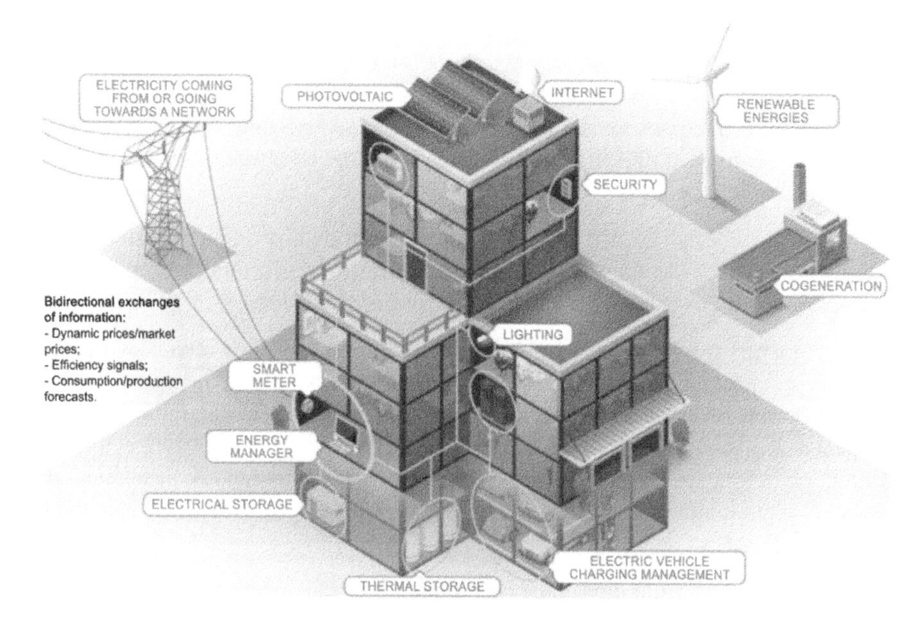

Figure 3.12. *Smart building connected to a smart electric network*[4]

A smart building integrates a series of technical equipment, which creates the interface between:

– all or part of weather conditions (heat, brightness, air quality, wind);

– all or part of the energy-consuming equipment that is used in the building (heating, air conditioning, ventilation, lighting, electric vehicle), local energy production and storage;

– all or part of the spaces and times when the building is in use, whether for residential purposes or activity on business premises.

The intelligence level of a building is not standard. It varies according to the range of services provided by the technical equipment within it. Several dimensions are used to evaluate this intelligence which is developed (ideally) for the dual purpose of maximizing energy efficiency and comfort of use. There are four of them [ROB 19]:

– *Automation*: according to a standard scheme which involves taking a measurement or making a time-related adjustment, a technical tool (e.g. a sensor)

4 Available at: http://www.objetconnecte.com/batiments-intelligents-marche-iot/.

will trigger an action to adjust something (more or less heat, more or less light, more or less ventilation). For example, light sensors can control the switching on of lights or the activation of blinds according to sunlight and the time. Implementing such automations in a domestic setting is usually presented under the term home automation. Actions can be triggered through a dialogue between devices, with or without human intervention.

– *Centralization*: all or part of the device is controlled by a single system, such as a thermostat for heating housing. Exceptions to this centralization rule are however frequent: for example, thermostatic valves on radiators.

– *Regulation*: this generic term covers all actions, whether technical or not, automatic or not, which are intended to adjust the demand for energy and/or the level of comfort of users according to their needs. For example, a window that can be opened is a controlled device in its own right. In the same way, regulation, information about consumption or awareness-raising through eco-gestures are modes of regulating (or modification, to put it another way) the socio-technical conditions of energy use.

– *Information*: consumption must be monitored as closely as possible, that is to say, over years, months, weeks, days, hours, according to equipment, etc. The challenge is to make the link between technical devices, consumer devices, ranges of use and actual consumption. The information available is not sufficient to understand consumption practices, rather simply the times and sometimes the intensity of use (e.g. an oven which is more or less hot).

The first smart socio-technical demonstrator building at the Université Catholique de Lille, called the Rizomm, is illustrated in Figure 3.13. Built in several stages since 1956, hosting the masters of the Faculty of Management, Economics and Science, the Rizomm is today a low-energy building, self-producing and self-consuming electricity, with the following characteristics:

– exterior insulation and replacement of exterior joinery;

– terracotta sunblocker, awnings and external rolling shutters which limit direct sunlight in summer and increase user comfort;

– ventilation system with energy recovery;

– LED lighting with high luminous efficiency and automatic management of lighting by detecting the presence and the supply of natural light;

– a photovoltaic roof and local production plant: 1,200 m^2 of photovoltaic panels allow the Rizomm to produce and consume its own energy.

Figure 3.13. *First demonstrable smart building at the Université Catholique de Lille, called the Rizomm (source: UCL Media Lab)*

The technologies deployed are designed so that occupants retain some control over how they manage their own comfort levels:

– individual adjustment boxes which allow the temperature and brightness of spaces to be fine-tuned within pre-programmed tolerance margins;

– energy coaching application for a smartphone.

An energy and performance building management system supports users when getting started with new systems. The Rizomm makes it possible to test the energy solutions of tomorrow on a real scale using an interdisciplinary approach, which has been the subject of socio-technical research and training (via a research study on the behavior of users, use of data, etc.). It allows users to be fully involved in energy performance. Chapter 5 of this book deals with the socio-technical dimension of smart buildings.

The second building in the university that has been transformed into a smart building is that of the School of Higher Engineering Studies (HEI) which is integrated today with the Junia school. A digital image (BIM for Building Information Modeling) of the historic building first built in 1885 is shown in Figure 3.14. The objectives are as follows:

– to improve the energy performance and reduce the carbon impact of the building, among other things, by making it a self-producer of photovoltaic energy, and by involving users in order to reduce energy consumption by nearly 10% via by behavior changes;

– improve the operating performance of the building via the technical building management infrastructure (BMI);

– to improve the services offered to the various users (pupils, staff, visitors, operators, etc.).

Figure 3.14. *Historic HEI building in Junia transforming into a smart building (source: Grégory Vangreveninge)*

The transformation of HEI buildings into smart buildings[5] makes it possible to centralize all information related to the buildings (energy consumption, space temperatures, CO_2 concentration in classrooms, etc.), but also to control and ensure the operation of equipment such as heating substations, air handling units, circulation lighting as well as the heating valves and the ventilation damper within all the rooms in the buildings.

Since the HEI buildings were renovated in 2015, the transformation into a smart building did not have the intention to replace the equipment that was previously installed, but rather to improve their use in order to reduce their energy consumption while guaranteeing that users are comfortable.

In order to be able to optimize the periods during which the different spaces are heated, presence sensors have been integrated in all the rooms of the buildings to be able to adapt the temperature set points according to the actual occupancy as well as connected window handles which allow the heating to be switched off if users wish to open them. Connection between business software (e.g. Enterprise Resource Planning or ERP, ticketing to manage task prioritization, etc.) and WiFi terminals has been undertaken.

---

5 Details on the HEI smart building are provided by Grégory Vangreveninge, technical manager of demonstrators at Junia.

The monitoring of buildings and their environment has been supplemented by additional functionalities which concern air quality and meteorology, through the installation of a meteorological station and outdoor air quality sensors and interior. Eight pollutants are thus monitored simultaneously (CO_2, CO, VOC, NO_2/O_3, PM 1, PM 2.5 and PM 10), as well as indoor and outdoor humidity levels, and outdoor noise.

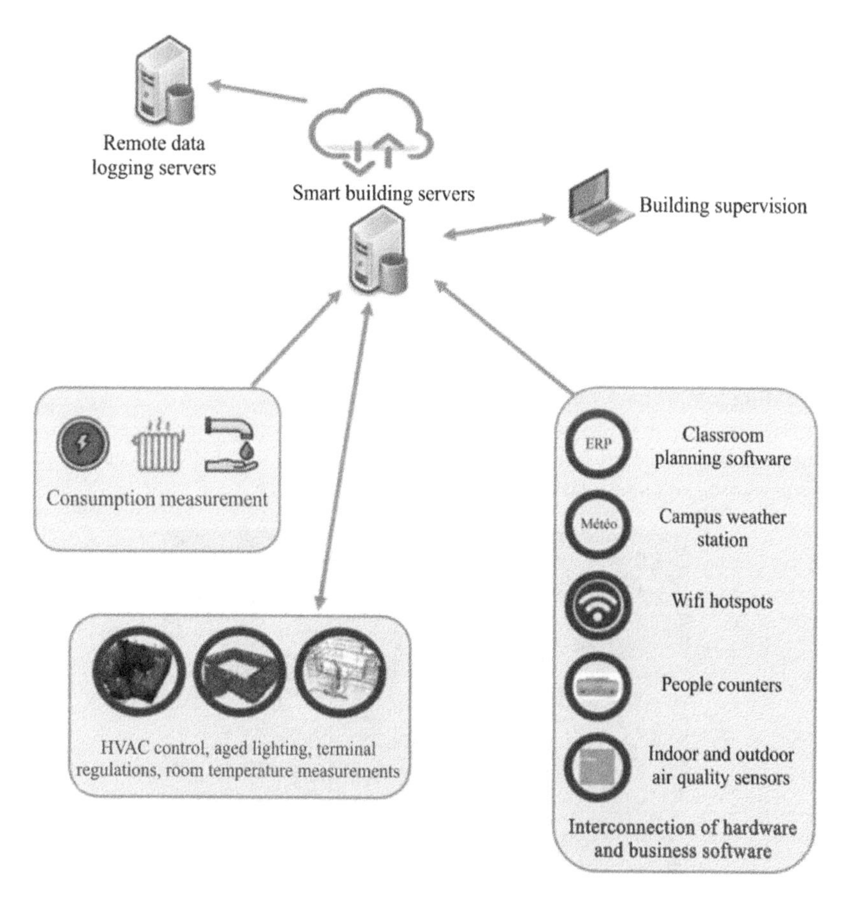

Figure 3.15. *Interaction between the different equipment making it possible to transform the HEI building into a smart building (source: Grégory Vangreveninge)*

Because of this new equipment, new user comfort management algorithms will be developed. This is because the recommendations and regulatory obligations related to the temperature instructions for example were not adapted according to use of certain spaces. Regulations require tertiary heating systems to be sized in order to provide a temperature in these spaces of 19°C ± 2°C. The temperatures felt vary according to different individual parameters (sex, physique, state of health, static work or not, etc.), but also meteorological and situational ones (air humidity, wind direction and speed, exposure in the sun, etc.). The combination of these meteorological and situational parameters makes it possible to apply temperature instructions that are more suited to the actual uses of our spaces.

Devices that count the number of people at all the entrances and exits of the buildings initially make it possible to determine their attendance. Because of these counters, as well as the presence detectors in each room and the data transmitted by the WiFi terminals located in the building, the concentrations of people in the spaces can be estimated while respecting the anonymity of the occupants. These data allow one to better optimize the energy management algorithms.

The performance of a smart building in terms of the service it delivers to its users, operators and managers is essentially a function of the interconnection of the building with its entire internal and external ecosystem. The building is thus able to collect and use all the data at its disposal to provide new services to its users. Figure 3.15 shows the interaction between the different equipment to transform the HEI building into a smart building.

The Rizomm and HEI buildings are two components of a block of demonstrator buildings shown schematically (BIM model) in Figure 3.16.

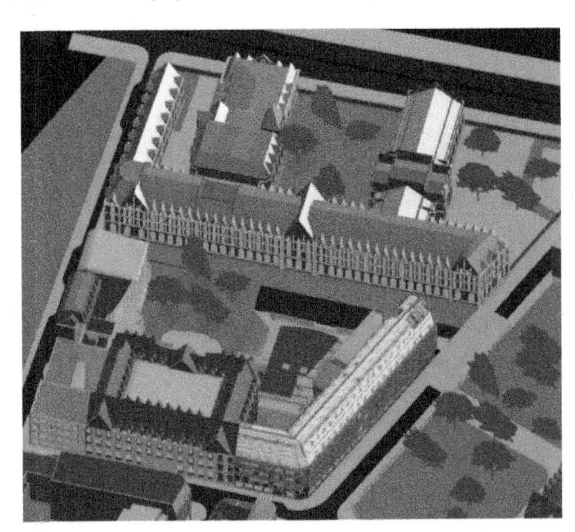

Figure 3.16. *Block of demonstrator buildings of the Université Catholique de Lille. Bottom right, the Rizomm building; bottom middle and left, the HEI building; and behind the academic hotel from 1877 (source: Foundation)*

3.4.4.3. *Smart grid and self-consumption*

Like buildings, energy networks are equipped with sensors that collect a lot of information, systems to transmit and manage this information, as well as the ability to act on different components of the energy system. This is particularly true for the electricity grid, which is increasingly integrating intermittent renewable energies (wind, photovoltaic, small hydro), as well as new uses such as the electric vehicle. The network must adapt to this new context, increasing its intelligence level so that each actor, producer and consumer contributes to this adaptation considering the energy efficiency objectives: reduce carbon emissions and make economic gains. The increasing storage of electrical energy offers new exploitation possibilities, which make it possible to shift away from random renewable productions, and to postpone and modulate consumption levels. The development of renewable energy sources encourages local production which is close to the loads that need to be supplied and therefore local consumption levels. This is particularly true for energy sources whose primary source varies independently of demand, without a natural form of storage (e.g. in hydraulic or fuel form), and whose production forecasts are marred by errors, as is the case of wind power and photovoltaics. In this context, self-production and self-consumption of renewable energy are defined as follows [ROB 19]:

– self-production is the share of total consumption that is ensured by local renewable energies, over a day, or more generally calculated over a year;

– self-consumption is the share of renewable production consumed in real time, over a day, or more generally calculated over a year.

Because of a photovoltaic installation of 150 kW peak installed on the Rizomm building, which is illustrated in Figure 3.17, an "intelligent" photovoltaic self-consumption system which is in accordance with the needs of the manager of the public distribution network has been the subject of experimentation within the framework of the So MEL, So Connected project[6]. These experiments and their potential extensions will be developed in Robyns et al. [ROB forthcoming]. This photovoltaic power plant helps to self-produce 9% of the energy consumed by the Institut Catholique de Lille buildings located in the demonstrator block over 1 year, for a self-consumption level of 98.2%.

The electrical installation of the first local smart grid demonstrator block of buildings which integrates photovoltaic production, electrical energy storage (250 kWh,

---

6 Available at: http://www.lillemetropole.fr/mel/institution/competences/energie/so-mel-soconected.html.

40 kW charging and 80 kW discharging) and controllable charging stations for electric vehicles (six 22 kW charging points) is shown schematically in Figure 3.18. Controlling the storage and charging of electric vehicles brings a level of flexibility to the electrical management of the block of buildings. In fact, it can absorb energy and send it back to be stored in the event of excess production renewable energy in relation to consumption needs, or according to the cost and the carbon rate of the energy supplied via the public energy distribution network, which is managed by the distribution system operator (DSO), for example.

Figure 3.17. *Photovoltaic installation of the Rizomm demonstrator building (source: UCL Media Lab)*

As part of the So MEL, So Connected project, an economic assessment of the self-consumption of photovoltaic energy was carried out with a view to identifying an economic model for the university. The business model of a company presents the distribution of activity and, in particular, the origin of its income in order to generate profitability. What is more, allowing players to design their business model requires bringing together basic elements[7] [ROB 19]:

– an identification of the players concerned and an understanding of their interest/risk in the context of the new activity in a new context, for example, that of smart electricity grids;

– a cost-benefit calculation on a consistent scope;

– once the scope of the calculation has been defined, an identification of cost items and sources of gains along the new value chain;

– an estimate of the financial implications of these gain and cost items;

– identification of potential funding sources.

The photovoltaic plant which equips the Rizomm building of the Université Catholique de Lille (150kWp, 140MWh/year) was designed with the aim of maximizing the production of electricity by maximizing its surface area (1,200 m^2), and by using high efficiency monocrystalline photovoltaic cells [ROB 21]. The warranty of the plant is 25 years. A technico-economic analysis was carried out considering life scenarios of 20 and 40 years. Indeed, feedback from photovoltaics in the North of France indicates that a lifetime of 40 years for monocrystalline cells is not unrealistic due to the lower average temperatures compared to the South. This is in the knowledge that what reduces life is the loss of cell efficiency over time, which also decreases as cell temperatures rise. The self-consumption rate by the Institut Catholique de Lille is estimated to be at 98.2% and the self-production rate is estimated to be at 9%.

A few working hypotheses had to be made:

– replacement of the inverter that drive the solar panels every 10 years at a constant cost, and annual cleaning of these panels;

– photovoltaic yield assumed to be constant (while it decreases over time) and value of the kWh not purchased via the distribution network assumed to be constant (while its value will increase over time) – one makes the assumption that these two variations will approximately compensate each other, although it is likely that the cost of electricity will increase more rapidly than the photovoltaic yield will decrease;

– it is assumed that self-consumption will not be subject to a specific tax. If no energy is returned to the public distribution network, there is no additional cost incurred by this connection.

7 Available at: www.smartgrid-cre.fr.

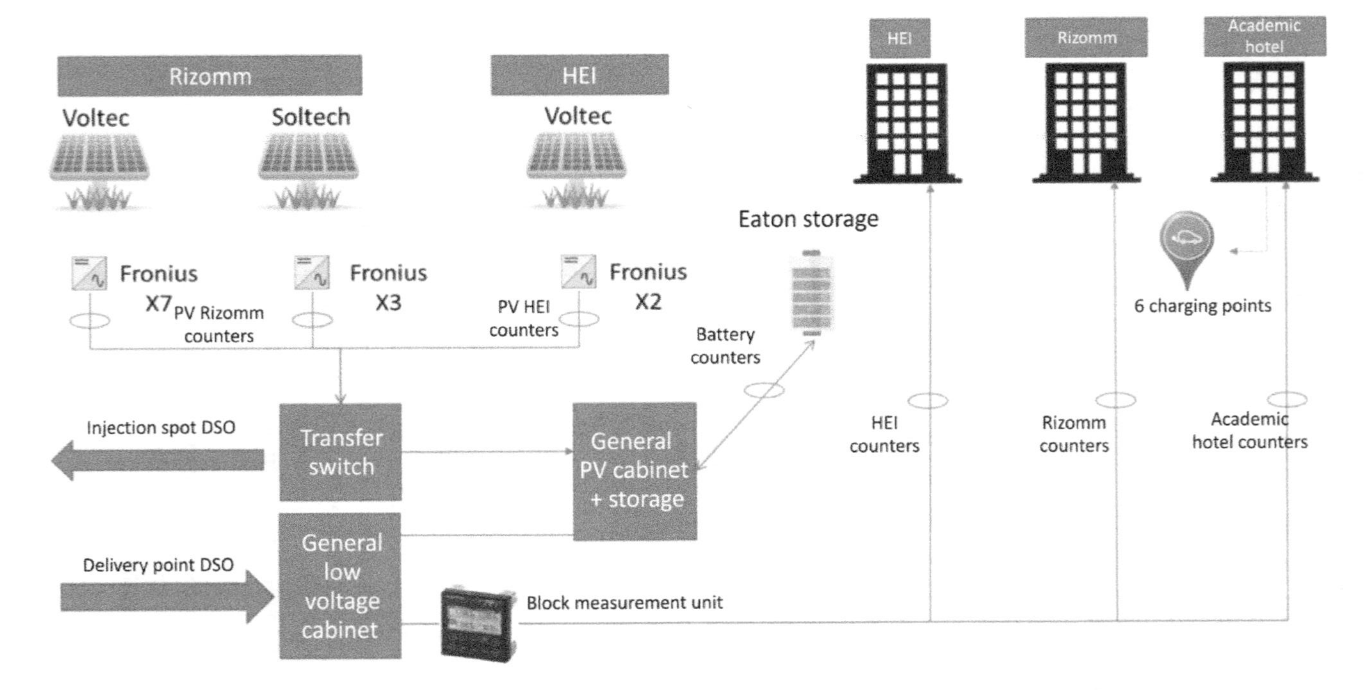

Figure 3.18. *Electrical installation of the first local smart grid demonstrator block of buildings integrating photovoltaic production, storage of electrical energy and controllable charging stations for electric vehicles (source: Grégory Vangreveninge)*

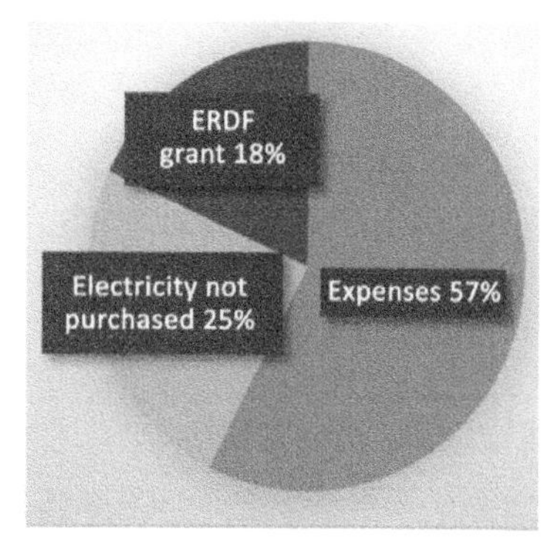

a)

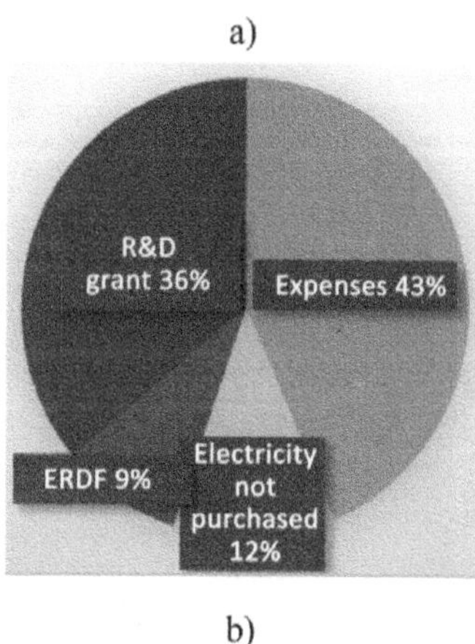

b)

Figure 3.19. *Economic assessment of photovoltaic self-consumption of the Rizomm building over 20 years without adding new values (a) and with adding new values (b)*

Figures 3.19 and 3.20 present the economic balance of photovoltaic self-consumption in the Rizomm building, respectively, over 20 years and 40 years, without adding new values (a) and with adding new values (b).

The charges over 20 years are excluding the added values of €654.13k, and with added values of €1,034.13k. The charges over 40 years are excluding the added values of €866.13k, and with the added values €1,246.13k.

Among the products, an ERDF grant (European Regional Development Fund) plays an important role; it is possible to say that it covers services that are provided by the university to society via initial training, openness to the neighborhood and its links with communities, which are difficult to quantify financially.

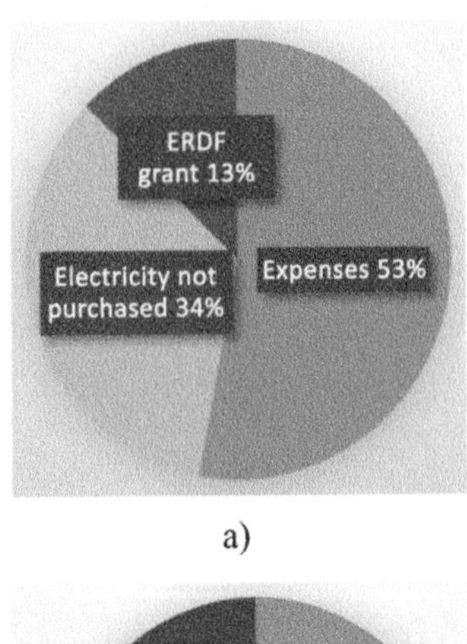

a)

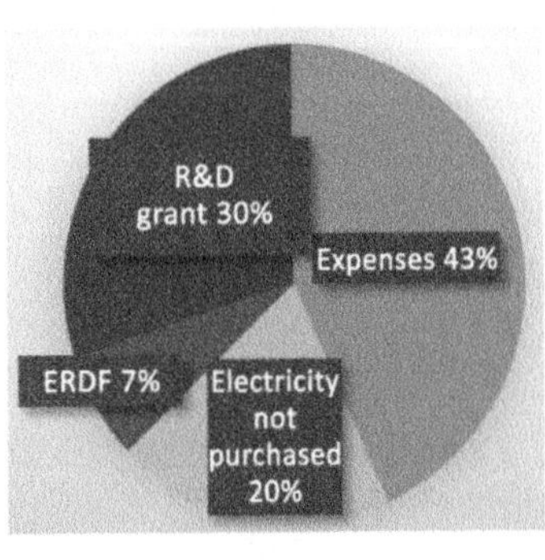

b)

Figure 3.20. *Economic assessment of photovoltaic self-consumption of the Rizomm building over 40 years without adding new values (a) and with adding new values (b)*

The difference between the revenues and expenses remains negative (expenses greater than 50%) with no new additional values added, but it decreases over the duration of the scenario (expenses decrease from 57% to 53% between Figures 3.19(a) and 3.20(a)).

Balancing the economic model will be achieved through several potential services:

– injection into the distribution network at the request of the DSO;

– removal of the distribution network;

– voltage adjustment at the point of connection to the distribution network;

– renewable energy charging of electric vehicles;

– charging electric vehicles by limiting distribution power (reduction of subscribed power);

– test/demonstrator site;

– research and development projects eligible for subsidies;

– specific ongoing training in photovoltaics and self-consumption;

– offer of services after evaluating self-consumption potential.

In the first year of operating the photovoltaic power plant, the university obtained grants for two research and development projects, one national (So MEL, So Connected subsidized by Ademe) and the other European (H2020 e-balance+), which means that the economic balance sheet of this self-consumption plant has become positive. The share of expenses has dropped to 43% in both scenarios. It should be noted that the subsidized projects also induce expenses which are taken into account in these balance sheets.

A second block of buildings, from the Institut catholique des arts et métiers (ICAM), located 500 m from the historic block, is also equipped with a local smart grid equipped with self-consumption photovoltaics, electrical energy storage energy and charging stations for electric vehicles. Throughout the Vauban campus, 26 electric vehicle charging points have been installed with the aim of making them effectively energy efficient.

3.4.4.4. *Information system and Internet convergence*

New modes of energy production, consumption and management involve the implementation of an Internet energy management system. As buildings present the main challenges in the approach, it will be necessary to associate an Internet of Things network that integrates the latest innovations in building modeling or BIM. Finally, our desire to integrate users into the heart of the system leads to the association of an Internet of people or the ubiquitous Internet which interacts with the global management system (via applications on smartphones, tablets or any other adjustment or communication device available to users).

An information system that is dedicated to Live TREE developments and experiments has been set up in order to acquire a means of storing and processing important data, with a concern for the confidentiality of these data and the protection of the privacy of all the actors involved in the experiments. This information system in the historic block is schematized in Figure 3.21.

The forecast of photovoltaic production is carried out in the medium term (15 min to 6 h) using satellite weather data, and in the short term (2 to 15 minutes) by means of a camera that monitors the sky (Figure 3.25). Long-term forecasts (those over several days) related to the production and consumption of buildings are made using artificial intelligence tools such as neural networks.

These forecasts are important for optimizing self-consumption, better managing storage, charging electric vehicles and managing interactions with the public distribution network. These forecasts must be supplemented by weather forecasts and room occupancy forecasts (the historic block can accommodate 5,000 people, students, staff, visitors, at peak times). These influence consumption and should be complemented by forecasts regarding uses and behavior shown by those who use the building.

An experimental wireless communication network with relay sensors is deployed in conjunction with the conventional wired network, with the aim of comparing different modes of communication and ensuring that this communication becomes redundant, which is a crucial factor in the management of all systems which manage to become *smart*.

An energy control center (schematized at the top left of Figure 3.21) as well as smart buildings have been developed to collect, manage and store the data retrieved from electrical equipment and buildings, in order to process consumption and production forecasts, and to control photovoltaic sources, storage and controllable loads such as electric vehicle loads.

Figure 3.22 shows part of the control center which is equipped with a SCADA (Supervisory Control and Data Acquisition) system. Figure 3.23 shows the first two blocks of demonstration buildings (the historic block with Junia and the faculties, and the ICAM block) equipped with photovoltaic power plants, storage systems and electric vehicle charging stations seen from the SCADA system. Figure 3.24 shows the consumption levels of the four buildings and the production of the two photovoltaic power plants of the historic block during a week in March 2022. Seen from the SCADA system, Figure 3.25 shows on the left a view of the sky from the camera, with a very short-term forecast of photovoltaic production, and on the right, the weather and outdoor air quality data which are measured in real time.

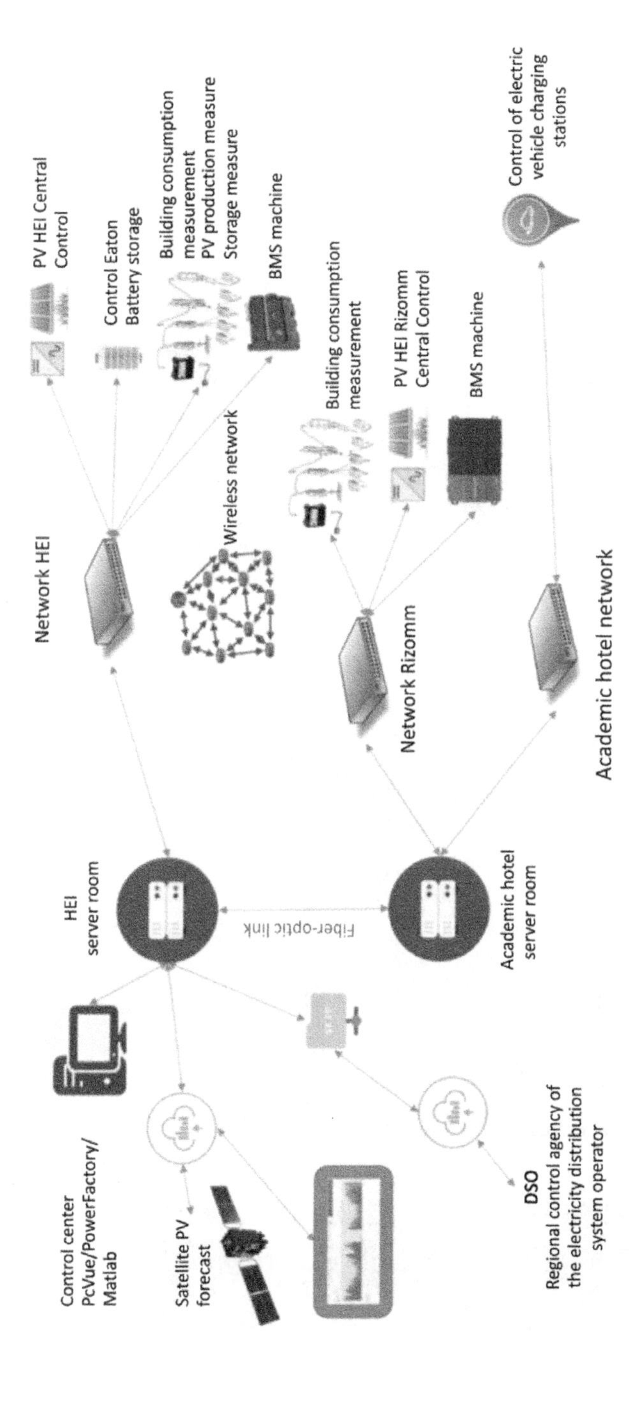

Figure 3.21. *Information system developed as part of the Live TREE program, and initially extended to the historic block (source: Grégory Vangreveninge)*

Figure 3.22. *Part of the control center equipped with a SCADA system (Supervisory Control and Data Acquisition)*

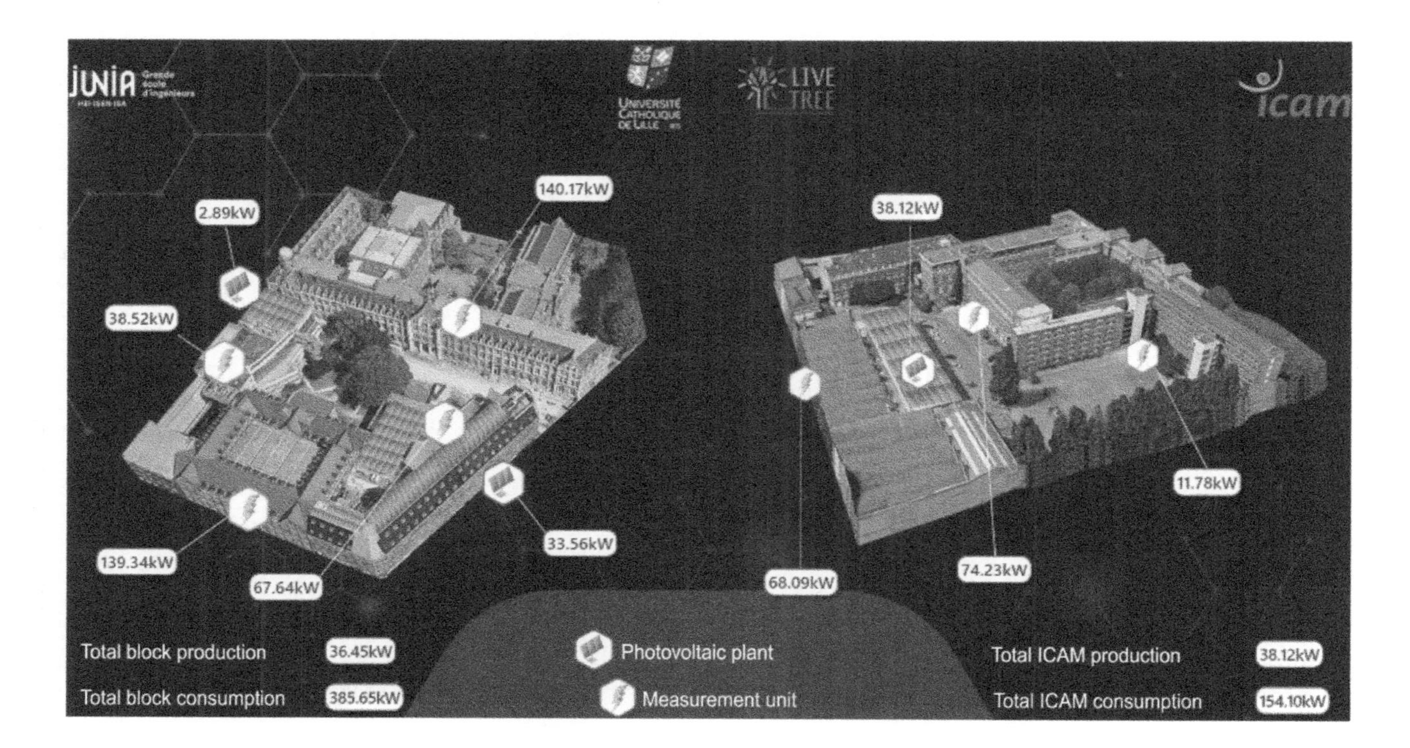

Figure 3.23. *The first two blocks of demonstrator buildings seen from the SCADA system (source: Anthony Aouad)*

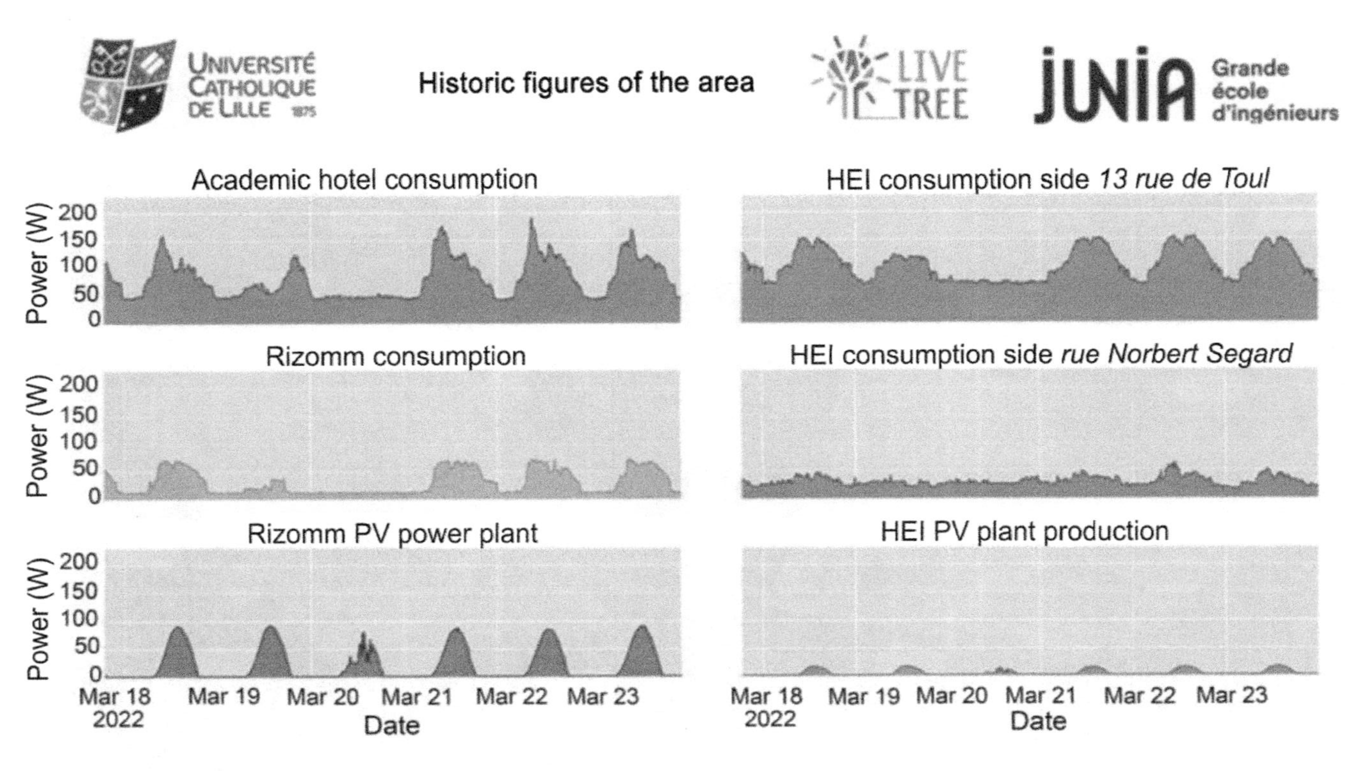

Figure 3.24. *Consumption of the four buildings and production of the two photovoltaic power plants in the first demonstrator block made up of the historic buildings of the university, the faculties and Junia during a week in March 2022 (source: Anthony Aouad)*

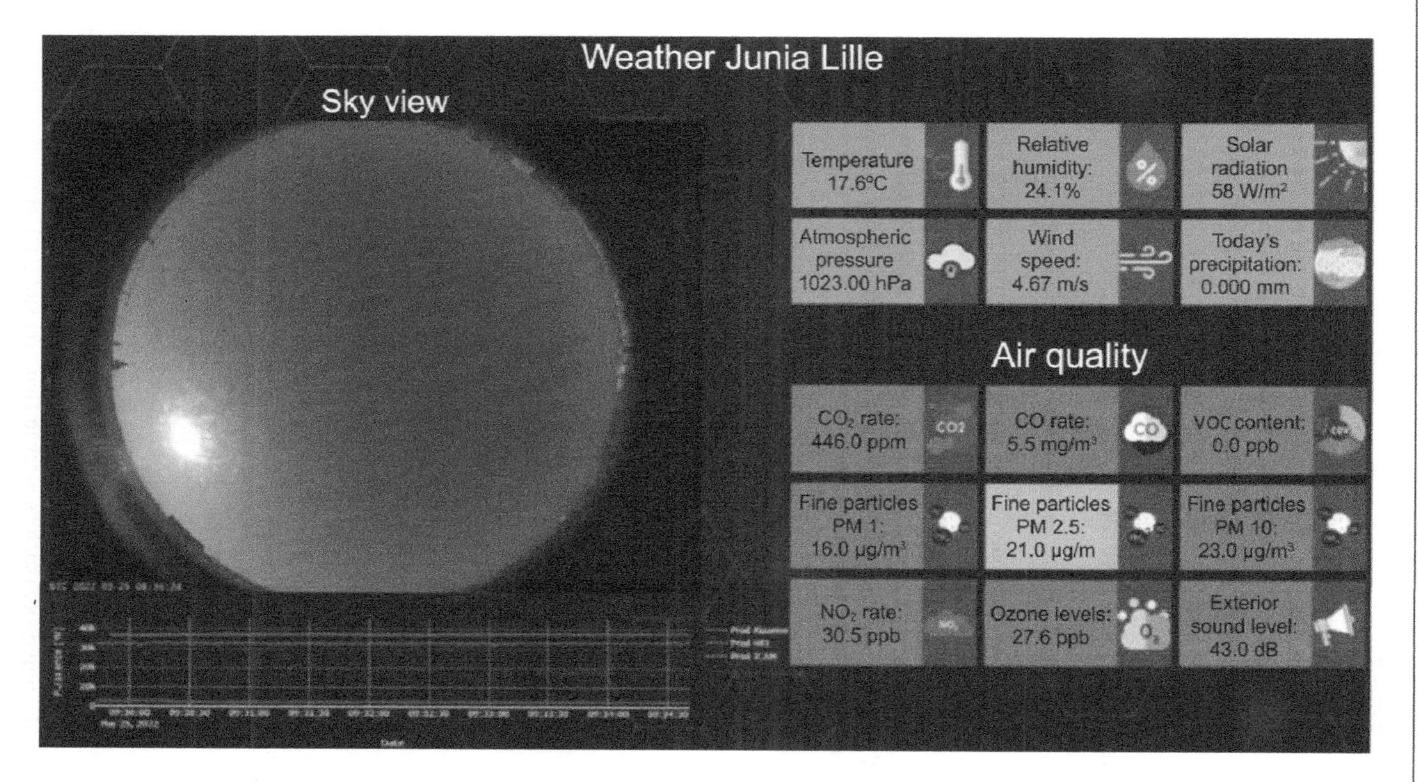

Figure 3.25. *From the SCADA system: on the left, a view of the sky from the very short-term photovoltaic production forecast camera, and on the right, weather and outdoor air quality data measured in real time (source: Anthony Aouad)*

Finally, the ambition of the Live TREE program in relation to the Internet of Everything is to implement and experiment with the convergence between the Internet of energy, the Internet of things including BIM, and the Internet of people or the ubiquitous Internet. This ambition must come with a profound questioning of ethical, sociological, legal, economic matters, etc., which associate the human and social sciences and the engineering sciences.

3.4.5. *Campus and zen district*

3.4.5.1. *An integrated urban campus open to its neighborhood*

The Université Catholique de Lille is located in an urban district near the center of Lille. Its buildings are scattered throughout the district and form the blocks that are shown in Figure 3.26. When launching its Live TREE program in 2013, the Université Catholique de Lille wanted to involve the Vauban-Esquermes district of Lille, its inhabitants and all its living forces: associations, schools, businesses and housing organizations and local authorities.

Figure 3.26. *Distribution of university buildings, shown in images, in the Vauban district of Lille. The historic area is located to the right of the center of the image (source: Studiographic ICL)*

Opening the buildings up to the neighborhood implicates issues related to mobility and the living environment within the campus. More specifically, the development of public spaces, the (semi-)pedestrianization and greening of roads, the possibility of giving residents access to technical mobility solutions that are developed by the university (such as electric charging stations), etc.

The ambition is to make Vauban-Esquermes a *living lab* on issues of living environment, mobility and integrating the university into the city, or into the reference district of Lille. This would be made possible due to:

– the construction of a Master Plan, or an action plan for a communal living environment, called zen campus 2025 (ZEN is also the acronym in French for *Zero Net Emission*);

– actions and experiments: treating the collective spaces (private and public) within the campus as demonstrators for the rest of the district, and as areas that can inform and make inhabitants aware.

3.4.5.2. *Green mobility*

According to Ademe, the company travel plan (CTP) is an approach aimed at tackling the problem of all travel related to an organization in a comprehensive and integrated manner. Concrete measures are thus taken to rationalize how users of an employment or activity site move about and to develop modes of transport that are more respectful of the environment. The CTP is concerned with how employees move around as well as other users of the site (customers, visitors, delivery people, trainees, etc.).

The university's first measures concerned developing the campus and the roads (creating cycle lanes, widening the sidewalks, opening a pedestrian axis transverse to the historic area, removing car parking spaces, street furniture, bicycle racks, etc.). This was alongside new services coordinated that were with the public collectivity (bus lines, bike rental stations), optimizing car park management, creating communication operations and trial offers.

Figure 3.27. *A self-service electric bike charging station at the Institut catholique des arts et métiers (ICAM)*

In 3 years, between 2006 and 2009, while campus users declared that they had a better understanding of policies in favor of mobility and that they had changed their behavior so that it was more multimodal, the modal share of cars dropped significantly (from 72% to 57% for employees; from 25% to 11% for students), CO_2 emissions decreased by 40% (from 7,257t to 4,391t) and particle emissions decreased by 45% (from 2.4 to 1.3 tons).

More recently, some university entities offer financial incentives (bicycle mileage allowance), a carpooling platform, self-service electric bicycles (Figure 3.27), or even electric and hybrid service cars and a service for charging electric vehicles. These latest developments are an integral part of smart grid experimentation on campus.

3.4.5.3. *Bringing nature into the city*

Vegetation offers freshness due to evaporation, while promoting biodiversity and offering a certain sense of seasonality to cities. Sitting on a lawn, being in direct contact with the grass, is not only a visual comfort, but also a psychological one. An interior green wall improves acoustic and thermal comfort and air quality and recreates the conditions for well-being in nature, etc.

It is urgent to bring nature back to the city, so that we can better resist climate change, bring back biodiversity, including birds which are migrating from our neighborhoods, bring back local urban agriculture favoring short economic circuits, bring back more serenity, trap a little CO_2, etc. [UNI 19a].

A green interior wall, like the one shown in Figure 3.28, contributes to the absorption of noise for more comfortable acoustic levels. It also has an impact on thermal comfort, so that temperature contrasts can be reduced because of insulation being reinforced. This process works both to fight against extreme heat through the process of evapotranspiration and to lessen the effects of the cold. It lowers the energy consumption of the buildings and walls it covers. We spend on average 80%–90% of our time in closed spaces and it turns out that indoor air is much more polluted than outdoor air. Paints, glues, varnishes, solvents in houses or buildings give off toxic volatile organic compounds (VOCs). Having plants which are capable of filtering and dissolving these VOCs will contribute to the depollution and sanitation of indoor air.

Working or living in a green environment brings people together. “Small-scale green spaces” in particular, turn out to have a positive impact on social cohesion. Neighborhoods with more green spaces are less subject to attacks, violence and vandalism.

Figure 3.28. *Interior green wall at the ESPAS-ESTICE school. A regulation of the plants has a positive impact on the air quality in a closed environment. This type of installation is being studied for confined spaces of the future such as those on planet Mars*

Due to growing urbanization and an increase in the population, combined with the desire to move away from over-intensive agriculture, the share of agriculture in cities and near cities is gradually increasing around the world. Internationally, the North American continent has long practiced urban agriculture, and other projects are now coming to fruition in Europe and Asia.

The theme of urban farming is emerging as a global issue in the face of the increasingly significant concentration of the world's population in cities: it is estimated that in 2050, 68% of the population will be urban, which will represent more than 6 billion people. Urban agriculture takes several forms: agriculture on the outskirts of cities, shared or community gardens, cultivation on the roofs of buildings, production in dedicated buildings and even a series of start-ups working in food tech which are growing in the city.

The Live TREE program acknowledges the fact that biodiversity is an ingredient of a productive city. It nurtures well-being, good air quality, social ties, new economies, positive environmental impacts and local food. The laboratory city is experimenting with the agricultural production methods of the future: producing in quantity, quality and locally in restricted spaces. An ambitious urban farm project is being carried out by the Junia school in a historic building, the Palais Rameau, in the heart of the Vauban district (Figure 3.29). The smart farming dimension of this project uses a lot of data and automated management in order to obtain local production inside a building, by optimizing water, energy and fertilizer resources, and by minimizing pesticides across all seasons.

Figure 3.29. *Palais Rameau in Lille, which hosts an experimental urban farm from Junia (photo credit: Willy Pulse, Atelier 9.81)*

3.4.6. *Involving the students*

To succeed in the energy and societal transition, future generations must be at the heart of this transformation, that is to say the 38,500 students of the university. One may achieve this by creating buzz around the topic which gives rise to student initiatives, as well as by having the means to support these initiatives.

The university encourages the creation of student associations, which are student initiatives. Many associations (37 in 2019) contribute to the sustainable development objectives on various themes: sustainable food and agriculture, new technologies and energy, solidarity and mutual aid, the zero-waste objective, recycling, environmental awareness and support for sustainable projects [UNI 19b].

By way of example, since 1992, the Hélios association, which was created by student engineers from the HEI school in Lille (now part of Junia), has produced four prototypes of solar vehicles over time, which have been tested during races around the world. Their latest prototype, Hélios 5, shown in Figure 3.30, goes from three to four wheels, which is more similar to traditional cars. The surface of the solar panels has been enlarged to 6 m^2, providing more power to the engine, and it can reach a speed of 100 km/h.

The circular economy has given rise to many student initiatives, whether for the reuse of furniture, eyeglass frames, the creation of new objects from old ones or the recycling of waste. The promotion of sustainable and local agriculture is also at the heart of many associations.

Figure 3.30. *Solar car prototype made by students at Junia (source: Association Hélios)*

To make students aware of the issue of sustainable energy production, bicycles that produce electrical energy to charge smartphones and tablets have been made available to them. The orientations of the Live TREE program aim for this program to become a breeding ground that allows students to develop their own sustainable development project, in addition to seeking to inform, raise awareness and develop ecological values.

In order to allow students to propose actions that contribute to the development of the university's carbon trajectory which aims for neutrality by 2050, the university is organizing a University Climate Convention in the autumn of 2022, made up mainly of students (100 to 150 in total). The idea is to imagine concrete actions and transformative visions of habits and behaviors, of ways of seeing and being in the world.

By way of example, students from the HEI engineering school proposed the pictogram shown in Figure 3.31, which could be similar to a nudge (see Chapter 6), which is aimed at encouraging people to turn off their computers and putting their peripheral devices on standby in order to reduce energy consumption and therefore the climate impact of digital technology.

The stakes and the speed of the transition are considerable. The solutions necessary to make it a success, whether they be technological, sociological, local linked to global, co-elaborative and not systematically imposed, may cause concern and discouragement. They will also raise many ethical questions. Addressing young people and reassuring new generations that there is hope and promise for the future, Pierre Giorgini proposes a four-dimensional ethics: "An ethics of action, which works towards and seeks out the 'good act' in order to move forward. An ethics of conviction, which aims to reintroduce values which are at the heart of action. An ethics of moderation, which constantly consistently aims to pave a more positive third path, which never ceases to prove difficult for people. Finally, an ethics of argumentation and rhetoric, which allows one to consistently avoid the simplistic and radical paths. We must open ourselves up to an incessant rise in consciousness, as an open process and not as the culmination of an indisputable truth. To do this, always exercise a legitimate distrust in totalizing, global solutions, because they always end up becoming totalitarian" [GIO 20].

Figure 3.31. *Pictogram encouraging energy savings and therefore a reduction in the climate impact of computers*

3.4.7. *Research*

To succeed in the energy and societal transition, the different disciplines of the university must be brought together, in order to develop interdisciplinary approaches between human and social sciences and engineering sciences, which is essential for a successful transition that everyone can feel a part of. Research is one of the major missions of a university, along with training. Each scientific field has developed its own methods and language which is shared by an often exclusive community. Dialogue between the sciences is therefore made more difficult. Researchers need to simplify their discourse to make it more accessible to people who are not experts in their discipline. Exchanges between disciplines are essential when it comes to vocabulary and methods in order to more precisely identify the possibilities for dialogue. Interdisciplinary research, which crosses disciplines, and transdisciplinary research, which exploits the expertise of at least two disciplines with equal value, must be encouraged and recognized by the authorities that direct and evaluate research. The effort it takes to cross between disciplines is an undertaking that takes time to be productive. So far, the evaluations of research activities have been essentially monodisciplinary. Building trust takes time as different disciplines need to be respected on an ethical basis.

Innovative and promising technologies have not seen the success that was expected because potential users have not accepted them. These technologies have been developed essentially using a technical logic without taking the interests and the acceptability of the users into account while designing, from the beginning. For example, we can cite Google Glass, a pair of glasses with augmented reality that can film people without their knowledge, and the smart electricity meter in France (Linky) which, from a detailed measurement of the consumption levels in a house, would make knowing the habits of its inhabitants possible. It cannot be claimed that any technological solution will be beneficial for the planet without integrating the sociological, psychological, ethical, economic and legal dimensions as far upstream as possible in design. That is to say, it is important to develop interdisciplinary approaches and support the change that is induced by the transition.

Interdisciplinary research which aims to develop full-scale, inhabited demonstrators can fuel "action-research", which can be defined as research for which a deliberate action has been taken to transform reality. It is research with a double objective: to transform reality and to produce knowledge concerning these transformations [WIK 20d].

The university develops interdisciplinary research on several subjects: sociological and ethical questions which are linked to the acceptance and involvement of users in smart buildings and smart grids, the importance of forecasting at all levels of production, consumption, economic and behavioral, modeling of the involvement of actors in a local electric smart grid with a view to it being supervised and able to self-consume within a local renewable energy community, the potentialities of smart buildings in interaction with smart grids and the contribution of demonstrators [DUR 20, DUR 21, ROB 19, ROB forthcoming, STE 20, STE 21].

This research work is, among other things, developed within the framework of two chairs of the Université Catholique de Lille: the exploratory chair for the transition held by the Faculty of Management, Economics and Sciences of the Institut Catholique de Lille, and the chair of smart buildings as nodes of smart grids supported by the Junia engineering school. These chairs are complemented by two complementary and structuring European projects for the Live TREE program: LIFE-MaPerEn and H2020 e-balance+.

The LIFE-MaPerEn project for "Participatory Management of Energy Performance" is supported by the Exploratory Chair for the Transition. Buildings are important in GHG emissions. To reduce energy consumption and GHG emissions, technological solutions are implemented: thermal renovation, development of renewable energies, energy storage, etc. However, the technological solutions deployed do not produce the expected effects, either because of technical hazards, or because they are not adapted or not well appropriated. The impact of these solutions is reduced by an increase in energy demand and new consumption habits (rebound effect); the impact of the behavior of building users is poorly known and generally underestimated. Aware of the limits of a techno-centric approach, various actors, communities, companies, universities and social housing organizations have started to develop actions to raise user awareness. These actions have their limits: any injunction toward users is experienced by some as constraints or perceived as initiatives that set out to decrease comfort; the actions are specific or experimental, without massification and with little funding; individuals do not perceive the impact of their behaviors and tend not to be reached by informational campaigns; and there is scant exchange of practices between the different sectors (residential, tertiary, university). In response to this observation, the Institut Catholique de Lille and Junia (Université Catholique de Lille) and the city of Lille and Lille Métropole Habitat (LMH) are seeking to set up a new system to raise awareness and increase governance that contributes to modifying the behavior of users of tertiary and residential buildings, in order to reduce energy consumption and GHG emissions. This is with a view to implementing smart energy networks. The Life-MaPerEn project is organized around the following three objectives:

– design collaborative energy performance management tools targeting users of tertiary and residential buildings;

– focus the management of energy performance on cooperation to encourage those who use these buildings to reduce energy consumption;

– consolidate and ensure the long-term governance and management of energy performance by integrating into guidelines for the climate policy of the city of Lille, the European Metropolis of Lille and other territories in France and Europe.

The H2020 e-balance+ project entitled "energy balancing and resilience solutions to create flexibility and new economic perspectives for the distribution network" brings together 15 business and academic partners in nine countries. It is coordinated by the company Cemosa in Spain, and the Université Catholique de Lille (Junia and Institut Catholique de Lille) is one of the project's four demonstration sites (based on the electrical and communication networks shown schematically in Figures 3.18 and 3.21). The e-balance+ project aims to:

– increase the energy flexibility of the distribution networks by modulating load consumption, local production and storage;

– forecast available flexibility;

– increase the resilience of the distribution network;

– design and test new service models for the electricity system (voltage and frequency control, etc. [ROB 15, ROB 19]) so as to promote new markets based on energy flexibility.

The system consists of units that implement algorithms to forecast and manage the flexibility available to encourage demand response programs, and to increase the capacity of the distribution network in order to avoid congestion and to optimize its use. The project is testing different flexibility solutions: electrical storage, V2G systems (vehicle-to-grid, or reversible electric vehicle charging [ROB 19]), hot water generator and systems connected to the IoT (Internet of Things).

3.5. Acknowledgments

The Live TREE program has received funding from the Hauts-de-France region, European funding from ERDF (European regional development fund), Life (MaPerEn project) and H2020 (e-balance+ project), and funding from the Université Catholoque de Lille Foundation. It received support and funding from the European Metropolis of Lille (So MEL, So Connected project funded by Ademe, and funding of Junia's Smart Buildings as nodes of Smart Grids chair), and support from the city of Lille. It also received funding under the State-Region Plan through Project Sunrise. Several companies support Live TREE developments.

Many people contribute to the development of the Live TREE program; the authors would like to thank them via this non-exhaustive list: Benoit Bourel, Yohann Rogez, Grégoire Destombes, Flovic Gosselin, Jacky Deboudt, Coline Zhagar, Fabienne Verhaeghe, Francis Deplancke, Loïc Aubrée, Anne-Sophie Loison, Grégory Vangreveninge, Jérôme Crunelle, Franck Chauvin, Johanne Gea, Christophe Saudemont, Dhaker Abbes, Khaled Almaksour, Anthony Aouad, Benoit Durillon, Arnaud Davigny, Julien Chamoin, Zohir Younsi, Nicolas Gouvy, Serkan Surkulu, Jad Nassar, Laure Dobigny, Gabriel Dorthe, David Doat, Aude Flant-Meunier, Marylinc Rousselle, Anne-Marie Michel, Marion Chivoret, Yannick Urbansky, Chloé Wyremblewsky, Léa Brunelle … as well as the management teams of the establishments at the Université Catholique de Lille involved in Live TREE, and in particular Patrick Scauflaire, President-Rector of the university since September 2020, and Pierre Giorgini, former President-Rector of the university who launched the dynamics of the Live TREE program at the end of 2013.

4

Smart Building Nodes in Smart Energy Networks: Components of a Smart City

4.1. Introduction

Energy is a fundamental issue of the transition. This is why energy networks are set to evolve strongly toward smart grids, just as buildings are set to evolve toward smart buildings. This chapter introduces the concept of smart buildings as nodes of smart grids. Within this, the aim is to position the users, operators and owners of buildings at the heart of an approach that involves modeling and dynamic supervision of buildings and blocks of mixed tertiary and residential buildings, integrating uses and actors, with a view to transforming them into intelligent nodes of a smart grid. This concept is a step toward the development of smart cities, raising a series of research questions (associating energy, buildings, transport, urban farms, digital technologies, participation of inhabitants, etc.) which will be discussed in this chapter.

As nodes of a smart energy network, smart buildings will become increasingly active participants in the energy ecosystem, beyond simple service mechanisms to the smart grid, which they can become an integral part of. The idea is to develop a real smart building vision, by considering networked buildings (buildings of different natures, energy coherent blocks, districts, etc.), effectively integrating all the technological dimensions and interests of users, operators and owners, and by positioning themselves as smart players in energy networks, integrating the full potential of digital platforms.

4.2. Smart buildings as nodes of smart grids

4.2.1. *Smart grids*

A smart grid is a communicating electrical grid that integrates digital information and communication technologies (ICT) in its operation. It makes it possible to establish interactions between electricity grids and the buildings to which they are connected, but also industries, transport systems, producers, consumers of various kinds, etc. [ROB 19].

Interest in such a network comes from the fact that its management becomes distributed and bidirectional, compatible with the production of distributed renewable energies (produced on site, with the capacity for self-production and self-consumption and in the form of a local renewable energy community). This allows daily peaks and troughs to smooth out, as well as pooling, etc. The network is at the heart of developing the infrastructures necessary for low-carbon energies, and will allow for a more efficient, economically viable and secure delivery of electricity.

This notion of smart grids based on electricity can be extended to smart networks for heating, cooling, gas and water.

Smart grids are based on the use of new technologies and are an integral part of implementing the Third Industrial Revolution proposed by Jeremy Rifkin [RIF 12], the deployment of which is underway in Hauts-de-France, Grand-Duchy of Luxembourg and the metropolis Rotterdam–The Hague. The essential complements for developing smart grids are the communicating meter (smart meter) and dedicated energy storage, or load modulation.

Some problems and obstacles exist when it comes to developing smart grids[1]:

– Significant consumer involvement is mandatory to see results: in fact, residents and occupants must cooperate fully in order to optimize their consumption using the tools provided. The gains may therefore be difficult to perceive in the short term.

– The installation of such a network involves the collaboration of many different actors: local authorities, builders, promoters, energy suppliers, etc. Such partnerships are complex to set up, due to the multiplicity of stakeholders and the possible conflicts of interest that may arise.

– The introduction of digital technologies into the network entails risks of hacking and attacks, both for individuals and for businesses. IT protection measures should therefore be applied to all network nodes. It should also be noted that fears have been expressed regarding privacy if smart meters were to become generalized.

1 Available at: www.planbatimentdurable.fr [Accessed January 13, 2017].

– Until recently, the inability to redistribute the energy produced between buildings (it was compulsory in France to resell it to an energy operator) slows down any valuation of the "kW avoided". The French Energy Regulatory Commission (Commission de regulation de l'énergie française, CRE) along with the legislator is setting up mechanisms to gradually lift this brake following the recommendations of the European Union, with a view to allowing for the development of collective self-consumption and local renewable energy communities. This is also what is called the right to disconnect, therefore, energy autonomy[2] [ROB 19].

– The legal texts (in France for example) make no mention of energy storage devices. The latter are therefore used according to how they are used on the network, but this question often remains unclear.

The very principle of the smart grid leads one to consider the building as interdependent with the district, the networks and more generally the environment in which it is located. Smart grids are often linked to new buildings – and so one challenge is to extend their application to existing districts.

4.2.2. *The digital dimension*

Today, another component has gained importance. If we feel that the "intelligence" of energy networks cannot be conceived without computerized management, the new potential revealed by digital platforms, data management and the exponential development of treatment opens up promising prospects in this area as well.

Let us take the example of buildings. Until very recently, we mainly talked about energy-producing buildings. In the future, they will produce and consume energy, but they will also produce and consume data in ever greater quantities. And if we know how to establish a link between data and energy, one obviously considers developing new possibilities that entail significant efficiency gains.

This obviously implies commissioning smart meters that are capable of managing this information. Buildings, at least in part, would become the nodes of this new energy network.

If we want to go a little further, so-called *blockchain* technology also can be applied to the processing of information that will circulate on intelligent networks. This information storage and transmission technology, which functions without a central control body, could be applied perfectly and in a transparent manner to

2 Available at: www.smartgrid-cre.fr.

networks that implement energy exchanges from multiple distributed sources [ROB forthcoming, ETS 20, ETS 21].

All this requires a mesh network (that is not centralized) which is very different from the one we have today. Implementing it gradually would mean building 21st century infrastructure which would perhaps entail an effort of the same order as that which was made to build the infrastructures that were necessary after Second World War.

4.2.3. *Intersection between buildings and energy networks*

As nodes of the smart grid, smart buildings will become increasingly active participants in the energy ecosystem; beyond simple service to the smart grid, they are able to become an integral part of it.

Demand for electricity continues to grow around the world. Traditional approaches to meeting this demand would require significant resources for additional production.

Making the electricity grid "smarter" using information technology can offset part of the growth in demand by fine-tuning grid management. However, this is only half the solution. The other half of the solution must be to make buildings smarter and better integrate them into an extended power grid.

Currently, few tertiary, commercial and residential buildings have adequate infrastructure to participate in such a transformation. However, the deployment of smart meters will contribute to accelerating this transformation.

Smart buildings with intelligent network interactions between buildings, building blocks, districts, etc., will enable the implementation of innovative control technologies in order to save energy, reduce the cost of energy and reduce the constraints induced on the electrical grids due to the development of random renewable energies and electrical mobility. They integrate the needs of users and residents, ensuring that they feel comfortable and involved.

When properly connected to the power grid, a smart building or block of buildings can configure itself as a virtual backup station or as a virtual power plant (behaving like a conventional power plant [ROB 15, ROB 21]) so as to regulate the electrical grid and keep it in balance.

Each block of buildings potentially constitutes a node of the electricity grid and can be likened to a microgrid which provides multiple services to its owners and occupants, as well as to the energy networks [ROB 19]. A growing number of these microgrids will integrate energy generation with storage services. The operators of these "micro-enterprises", owners and companies, will be able to optimize the services and energy consumption in each node in relation with the managers of the energy networks and possible energy and service aggregators [MEM 16].

Figure 4.1 illustrates the concept of smart buildings as nodes of smart grids by combining the smart grid vision, that is to say, the multiple contributions for the electric network in this case, and the smart buildings vision which has benefits for buildings and blocks of buildings in terms of reducing CO_2 emissions, but also when it comes to quality of life for users, financial savings and even business. This concept will be a key component of future smart cities. The nodes are characterized by transformer stations and/or smart meters.

An aggregator will probably be needed to aggregate interactions with energy networks [ROB 19].

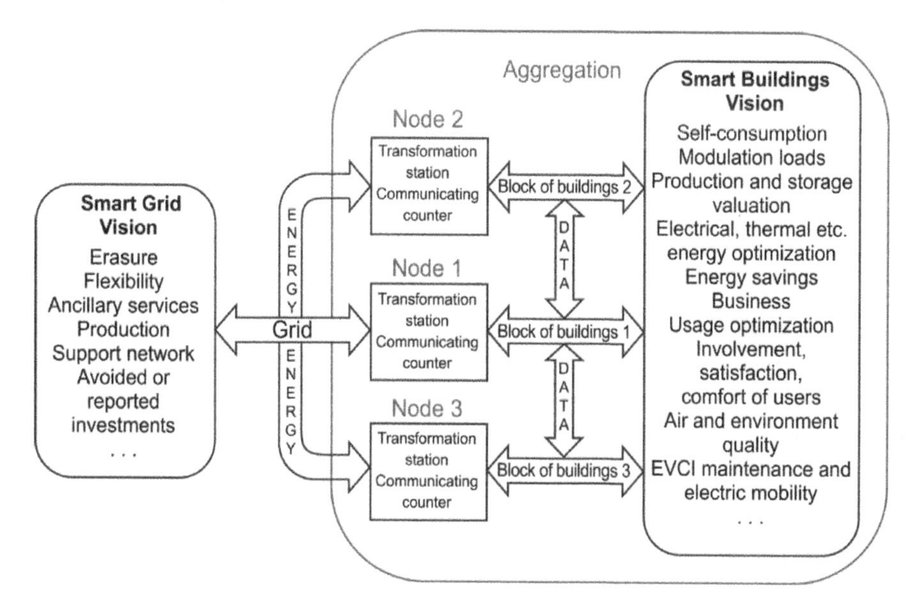

Figure 4.1. *Concept for smart buildings as nodes of smart grids (EVCI = electric vehicle charging infrastructure)*

4.2.4. *Transformation of buildings*

Transforming a conventional building into a smart building requires a reliable forecast of consumption levels. Consumption profiles are compared to the energy availability profile and an adjustment is possible if deviations occur. How building users are able to adapt their energy behavior is crucial to the success of smart buildings. Unlike in the past, when energy providers built expensive backup reserves into the system, consumers must play an active role in any future-oriented energy concept.

Building users or operators play a central role in establishing a realistic consumption forecast, since they are the only ones who know the processes in the buildings, and which energy-intensive processes may be reasonably delayed or advanced over time. This approach concerns electrical systems, but also thermal processes due to their inertia. One can allow a building to cool to a certain temperature before beginning to warm it up at the right time so as to maintain a comfortable zone. This can be implemented by taking the thermal storage capacities of the materials constituting the building envelope into account. Buildings are pre-heated using inexpensive energy at night, for example, so that they are ready for daytime use.

Consumers on the smart grid focus their attention on purchasing electricity when sufficient power is available at the lowest price. It may make sense to leave energy-intensive processes for a later time. For example, to heat an office to a comfortable temperature at 7 A.M., a typical building automation system might start heating or cooling the office an hour earlier. A “smart” solution can therefore be to start heating or cooling buildings hours in advance when electricity rates are low. However, in this case, more energy would sometimes be purchased, but at a lower price.

The best time to buy electricity is estimated using a logic software solution. The purchase of energy can be further optimized using correctly sized storage units, local cogeneration plants, solar panels and other structures for self-consumption. Consumers will balance their energy consumption according to the amount of local energy available.

Communication and coordination between consumption points and the network are crucial. Many current smart grid pilot projects can be significantly improved in this area. For example, some management systems now centralize shutting down all heat pumps within a given area during periods of high consumption in order to cap consumption peaks. If the pumps are then restarted all at once, consumption increases disproportionately due to the accumulated demand, producing another unnecessary peak. However, if a smart building is organized based on an incentive concept, it can respond flexibly to being stimulated and intelligently use the building as a holistic system instead of only involving individual predefined loads such as the heat pump. One thing is clear: without knowing the processes that go on in the building, uncoordinated external intervention always has a negative effect on comfort, safety and efficiency.

From these reflections, we can clearly see the importance of dynamically managing data collected in real time on a digital platform which processes the information using specific algorithms. There is much potential for optimization and therefore more efficiency needs to be developed.

Regarding the smart grid of the future, it is essential to consider participating smart buildings as autonomous and intelligent subsystems. Future solutions will aim to optimize the entire chain of all electrical and energy components of a smart building, from the power plant to individual lights in the workplace. An important aspect is to develop better storage methods. Local photovoltaic systems, for example, should produce energy as continuously as possible from a management point of view.

As the world's largest consumers of energy, buildings offer enormous savings potential. They can be made more efficient by integrating them into a smart grid; modern building automation systems form the basis of this approach. As smart, local participants on the grids, smart buildings play an important balancing role in the smart grid. Building operators and users thus benefit from economical, reliable, environmentally friendly and future-proof energy.

Deploying digital technologies within buildings to make them smarter and more efficient only makes sense if the buildings are basically low-consumption by design (ideally bioclimatic), with strong thermal insulation, very airtight, good performance of technical equipment (lighting, boiler, pump, etc.) and designed to obtain the best performance, the appropriate choice of heating methods (preferably of renewable origin), etc. Standards for defining building performance exist in each country and tend to evolve over time by becoming more restrictive.

4.3. Interdisciplinary R&D to move toward a *smart city*

4.3.1. *The smart city*

Today, cities occupy barely 2% of the surface of the globe, but they are home to 50% of the world's population, consume 75% of the energy produced and are the source of 80% of CO_2 emissions. Cities must therefore develop new, more energy-efficient services in different areas[3]:

3 Available at: www.smartgrid-cre.fr.

– Transport and smart mobility: various modes of transport, both individual (car, motorcycle, bicycle, walking) and collective (bus, metro, tram, taxi, etc.), will be integrated into a single efficient, easily accessible, affordable system, which is safe and environmentally friendly. This will allow the environmental footprint to reduce, while optimizing the use of urban space and offering city dwellers a wide range of mobility solutions that meet all their needs. In addition, the city of tomorrow will implement the latest public transport and electric mobility technologies [ROB 17].

– Development of a sustainable environment: cities will now increasingly undertake actions in the areas of waste and energy, for example, by developing eco-districts. In the field of energy, cities will strengthen their action in terms of energy efficiency (development of low-consumption public lighting, more efficient transport, etc.) and will set up local energy production systems (solar panels on the roofs of buildings, production of electricity from waste treatment, etc.) [ROB 19]. Whether for the purposes of creating energy via self-production and self-consumption of renewable energy, or for the production of products and food by means of urban farms, for example, short circuits will be developed in the city.

– Development of responsible urbanization and smart housing: the elevated value of real estate in city centers combined with the limited availability of land make current urbanization complex. Indeed, the model of urban sprawls that has prevailed up to this point, which are costly in terms of space, public facilities and energy, is no longer possible. Cities will put urban forms in place that will respect essential privacy, ensure sufficient sunlight, allow for changes and promote "living together". Buildings will also have to be smarter in order to facilitate and improve energy management, and even reduce consumption.

Citizens will play a central role in the city of tomorrow. They will no longer be considered as consumers of services, but as partners and stakeholders in its development. This new place will be granted to them because of the democratization of information which allows for more participation.

Several definitions of the smart city exist. Isam Shahrour summarizes these definitions by emphasizing that the smart city must be inclusive, collaborative and green [SHA 21].

4.3.2. *Interdisciplinary R&D*

Figure 4.2 summarizes the complementarity of R&D expertise (research and development), which is non-exhaustive. This will make it possible to develop interdisciplinary approaches which are adapted so as to contribute to the development of the smart city. Within the framework of the concept of smart buildings as nodes of smart grids which was introduced in this work, the first two axes, being smart grids and smart buildings, will be taken into consideration more specifically (this concept is developed in more detail in [ROB forthcoming]).

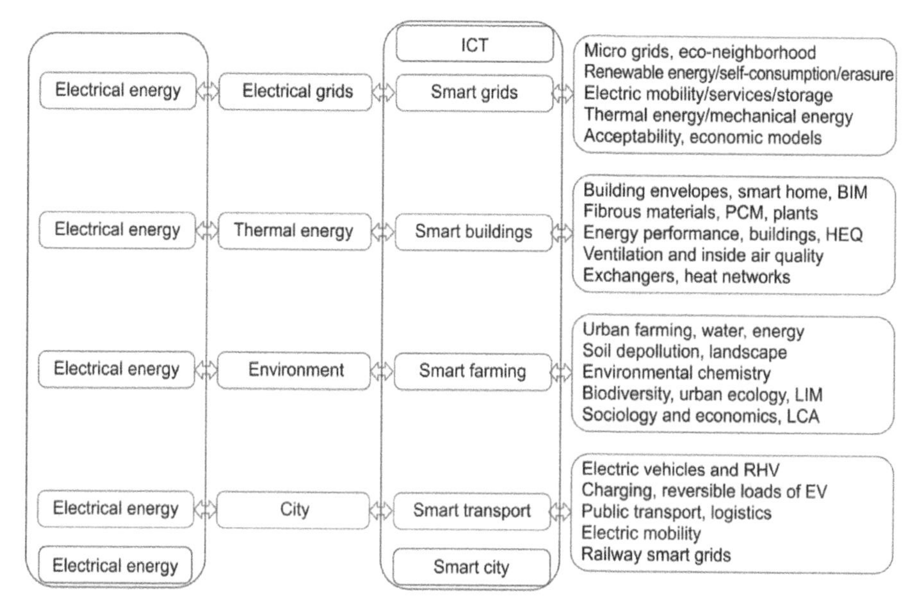

Figure 4.2. *Non-exhaustive R&D expertise that will make it possible to develop interdisciplinary approaches to adapt and contribute to the development of smart cities (BIM = Building Information Modeling, PCM = phase change materials, HEQ = high environmental quality, LIM = Landscape Information Modeling, LCA = lifecycle analysis, RHV = rechargeable hybrid vehicle, EV = electric vehicle)*

Smart farming, a component of urban farming, is the association of increased use of information which is collected and processed with diversified technologies. Farmers use it to produce and distribute agricultural products, as well as to regulate input supplies (water, fertilizer, energy, etc.). Smart farming can go as far as being a network of connected farms which exchange information allowing them to optimize their agricultural practices.

Smart farming is characterized by the use of technologies around three axes:

– data collection on farms with a view to optimize them (digitization);

– simplification of the operational work of farmers (robotization);

– a network of connected farms: the optimization of activities and interactions across the entire value chain.

By 2050, more than two-thirds of the world's population will live in cities. In order to feed it with our current agricultural techniques, one billion hectares of additional crops would be needed, the size of Canada. However, 80% of arable land has already been exploited. Moreover, extending cities encroaches on arable land. Urban agriculture or *urban farming* can provide a solution to this problem, but the total surface area of these outdoor plots will never be sufficient to cover future food needs. Vertical farming superimposes crops on several floors in the heart of cities, inside buildings. The *smart farming* dimension uses a lot of data and automated piloting in order to obtain local production inside a building, by optimizing water, energy and fertilizer resources and minimizing pesticides, across all seasons.

Smart transport, which is a key application of the Internet of Things, refers to the integrated application of modern technologies and management strategies into transport systems. These technologies aim to provide innovative services related to different modes of transport and traffic management, allowing users to be better informed and to make safer and "smarter" use of transport networks.

In 2010, the European Union defined Intelligent Transport Systems (ITS) as systems "in which information and communication technologies are applied in the field of road transport, including infrastructure, vehicles and users, as well as in traffic management and mobility management, and interfaces with other modes of transport".

Intelligent transport includes the use of several technologies: basic management systems such as car navigation; traffic light control systems; container management systems; automatic recognition for license plates or speed cameras to monitor applications, such as security video surveillance systems; and more advanced applications that incorporate live data and commentary that come from a number of other sources.

ITS technologies enable users to make better use of transportation networks. They also pave the way for the development of smarter infrastructure to meet future demands. The evolution of intelligent transportation systems is providing an increasing number of technological solutions for transportation managers who are looking to operate and maintain systems more efficiently and improve their performance. For example, a transport Internet which links trucks that transport freight would reduce the number of empty trucks by around 50% by optimizing the distribution of freight transport.

Figure 4.2 highlights the structuring nature of electrical energy and ICTs. The diversity of expertise highlighted by this figure underlines the need for an interdisciplinary approach to address this shift toward "smart systems" that can interact in order to respond in the direction of energy and societal transition, by integrating users and different actors who could potentially be smart users.

The concept of smart users will be discussed and questioned in the context of buildings becoming intelligent, or smart buildings, in Chapter 5, while Chapter 6 will ask the question of how we can influence individuals so that they play a part in energy and societal transition, among other things, via *smart* technologies designed to contribute to it. This obviously also raises ethical questions.

5

An Energy-Efficient Smart Building with or without the Cooperation of Its Occupants?

"An energy-efficient building is a technical object which is difficult to design, build, regulate, operate and occupy"

[BES 15]

5.1. Introduction

The harsh reality of climate change accentuates the urgency to find effective and rapid solutions to significantly reduce greenhouse gas (GHG) emissions. Among the emitting sectors, construction represents a significant share in all countries, so much so that it has become a main target for the national policies put in place, in particular concerning compliance with the objectives set by international agreements. In France, data show the share of the buildings to be at almost 45% of the final energy consumed, the residential sector taking around two-thirds and the tertiary sector the rest (around 30%).

The contribution of buildings to the reduction of GHG emissions, via a reduction in energy consumption, did not happen spontaneously within national territory. We know that the creation of thermal regulation (TR) as it is known dates back to 1974, in reaction to the oil crisis of 1973. It broke with a regulatory approach that was inherited from the post-war context, opting instead to put the emphasis on increasing the sanitary comfort of housing thanks to a generalization of hygiene points (toilet areas and amenities) and mains drainage. At that time, the main priority was to successfully reduce energy consumption, and therefore dependence on supplier countries, by mobilizing two approaches: regarding new construction works, it was about proclaiming a consumption reference per m² (225 kWhep/m²/year)[1]; and when it came to the entire stock (new or renovated in its state), setting an indoor set temperature point of 19°C[2].

The principle of thermal regulation which allows one to target several objectives at the same time (trade balance, energy dependence and geopolitics for that of 1974) will quickly impose itself as a major action lever for the public authorities. In line with the strategic challenges of the moment, successive thermal regulations will thus help to reduce the threshold of the desired levels of energy performance, to widen the perimeter of the buildings concerned and to embrace new challenges (following the influence of the Rio summit in 1992 and the protocol of Kyoto in 1997), such as reducing GHGs, undertaking the energy transition and encouraging the emergence of green growth.

Obligations for regulatory compliance and respect for consumption standards, which as we have said have become increasingly demanding, have revolutionized in 40 years the materials that are used within buildings, for construction engineering and for energy supply. This same generalized dynamic of innovation has embraced the digital sciences as well as artificial intelligence and has given rise to smart grids, smart buildings and smart cities.

The diversity of ways to apply this embedded intelligence in networks and buildings (by country, by construction sector and by operation) somewhat complicates a simple and universal definition of it. However, if we draw inspiration from the French context, which benefits from *world-leading* building companies, we can argue that smart buildings are "constructions that integrate measurement, detection and/or supervision equipment, designed with the aim of reducing energy consumption to a minimum, or even generating a positive balance in the event of local production, while ensuring a high level of quality when they are in use and a high level of service to their occupants".

Any definition will naturally need to emphasize the capacity of existing intelligence to reconcile, on the one hand, control over energy consumption which is already deemed to be economical in a constructive context, and on the other hand, a high quality of comfort and service to the benefit of the occupants, in their lives or work activities. However, the definition is limited in that it ignores the role attributed to these same occupants to achieve the set objectives. It is true that tackling this question is not straightforward. Depending on the operations in question, the energy-related procedures (programming, adjustments, etc.) may or

---

1 Kilowatt hour petrol equivalent, or kWhep in the following text.

2 The 19°C serves as a reference value (CIT = conventional indoor temperature) for forecast and regulatory calculations, one of the requirements of RT 2012.

may not be carried out at all times using technical tools alone (schedulers, detectors, probes, technical management of the building[3], supervisors, etc.) without any expected action on the part of the occupants, even if their presence cannot be detected by sensors or their needs are not expressed to the services offered (consultations, reservations, validation). The variable positioning that the occupants have modifies what is commonly called "the philosophy of the building": there are thus very technical smart buildings which exclude the occupants from the settings that are linked to energy (heat, light, occultation, ventilation, air conditioning), and others which do not. The information that is available today shows that the very technical philosophy largely dominates when it comes to operations that can be described as smart.

The observation made highlights how the technicist approach dominates when it comes to energy objectives. Naturally, we could sit back and say "why not", that what we are facing is a social production in its own right. However, when analyzing existing operations in depth, we can see problems related to energy performance, as well as a host of socio-technical assumptions. Regarding the energy issue, we are aware that internal intelligence does not allow the buildings in question to evade the significant trend we see of exceeding the energy objectives that are associated with thermal regulations, even though the decline in consumption is indisputable from one thermal regulation to another. The continual tightening of thermal regulations has therefore had positive effects on consumption. Yet, these effects have not been as significant as hoped and pose a problem in terms of compliance with commitments which have been made at national and international levels. On a socio-technical level, work being carried out linked to energy practices, forcefully or with the permission of inhabitants, undermines the presupposition of passive occupants that their needs and expectations will be met in terms of comfort, leisure and workload (whether domestic or not). Whatever the generation and the state of the building, we always note the existence of acts among the occupants which are intended to "build" the qualities of comfort and desired use (for a chosen moment, a precise duration or a given place) beyond the functioning of the building that they occupy. In any case, this tends to be the point of view of the people who decide to take action. In fact, the unanimous nature of this observation is at odds with the presupposition that there is a level of technical sufficiency which means that all objectives defined in terms of energy and quality of use can be met.

The name smart building refers to the authentic modification of the technical contents of a building, and therefore of how they function from a socio-technical perspective, considering the domination of a largely technicist logic. Under these conditions, the designation of so-called smart users which evokes users appears ambiguous, even paradoxical. Because the overriding tendency of the model is to

3 Technical management of the building (TBM).

ask nothing of the occupants, what intelligence are we really talking about, considering the fact that occupants are never really passive when making use of buildings in ways that are adapted to their needs in a given moment? The paradox between the technicality of smart buildings and the impossible neutrality of the occupants is at the heart of the socio-technical problem of this new construction model, which means that it deserves to be questioned.

Before going any further, it is essential to remember that the paradox being evoked is ultimately not uncommon. The emergence of smart buildings only changes the nature of the difficulties that construction models designers face when integrating the parameter of uses. However, does the smart buildings model more effectively overcome difficulties? This is the central socio-technical problem, which is addressed in this chapter in an explicit and overt manner, but which influences this work.

The analysis that is developed in this chapter benefits from feedback from two buildings of the Université Catholique de Lille which have been renovated and transformed into demonstrator smart building as part of the Live TREE program: the Rizomm building of the faculties and the HEI building of Junia (see section 3.4.4.2). They share the same academic vocation but hold socio-technical philosophies which are radically different. The older building is HEI, which dates back to 1885. The transformation of such old buildings into smart buildings is obviously a challenge.

5.2. Construction methods for energy performance

In order to question the problem that is posed by smart buildings, it is necessary to look at the evolution of the ways of designing and constructing buildings in a connected way, as well as the associated socio-technical problems. Three stages must be marked out: construction methods which are far from the scale of energy consumption, the time taken to search for energy performance and finally, the technical-functional reconciliation between energy production and consumption, which is made concrete by smart buildings.

5.2.1. *The time to satisfy basic needs*

Energy has always accompanied the history of mankind; it has always been a question of survival. Being provided with light, heat, food, protection, being able to transform matter, to be transported, etc., none of this would have been possible without energy. With the passing of time, it is not so much needs that have evolved as it is the available energy sources (wind, water, animal power, wood, oil, etc.) as well as the technical devices associated with these energy sources, which have sometimes had a more significant impact on ways of living.

In a simplified way and focused specifically on the issue of domestic buildings, energy consumption has long been related to the essential functions of heating, lighting and cooking food. The same energy consumption could also have concomitant functions, such as the energy generated by a chimney fire which warms, cooks and illuminates.

What is the sociotechnics of energy?

The sociotechnics of energy was born from the first use of an energy source. It corresponds to all the acts, devices and tricks imagined to obtain, improve, support, transport, make it last, control the cost, etc. of energy intake, with as little mental and physical effort as possible. Taking an interest in sociotechnics therefore means identifying, understanding the why and the how, as well as questioning the effectiveness and induced effects (costs, GHGs, exhaustion, etc.) of direct and indirect human interventions on objects related to energy. The social modalities of its implementation also conceal strong symbolic and power dimensions. This will not be discussed in this chapter.

Box 5.1. *Definition of the sociotechnics of energy*

Depending on the energy available, the climate, the local geography, household finances or housing, uses associated with energy have developed and diversified over time; studying them is a matter of socio-anthropology. A few simple illustrations help to make the outlines concrete:

– the family gathered around a fire, and living in a single room;

– the proximity maintained between livestock and the occupants of a dwelling to take advantage of the heat from an animal;

– the arrangement of carpets on the walls of lords to limit the flow of heat to the outside and the sensation of a cold wall;

– the layering of clothing to keep body heat in;

– the installation of thick curtains on the windows and rolls at the bottom of the doors;

– the introduction of previously heated objects to warm the sheets before sleeping (bed warmers) or for as long as possible during the night (hot water bottles, heated bricks, etc.);

– sleeping in the same bed with someone else to keep warm.

This selection of illustrations forms part of a simple technical context, which is centered on fire (the fireplace, the wood oven, the candles), due to the fact that it was the type of energy used for the longest time in the home. The diversity of uses makes it possible to introduce the first important notion, which is "energy intelligence". The concept encapsulates both the reflective exercise which is provoked by an energy issue and the practices devised to respond to it. Here, we are talking about intelligence in the literal sense, since it is necessary to find a solution to a problem. Reproducing a solution found before and/or elsewhere does not detract from the value of what is found. The important thing is that the desired result is achieved. Successfully repeating the sequence "problem-reflection-solution" increases the range of solutions that are available. It is the basis of energy intelligence, which becomes a crucial element in intergenerational experiences, a real habit. The following quotation emphasizes the importance of the technical and environmental context because it demonstrates the validity of the cognitive exercise and its effectiveness:

> Modes of living are not subject exclusively to the domination of technical systems, even if these remain strongly prescriptive. The domestic space can just as much be described as the place where skills are built, which we call "energy intelligence". These skills confirm the involvement of the individual, through their body in the construction of their living space [SUB 09].

Without going too far, we can say that this period, which had little to no concern for the impact of energy consumption, lasted until the early 1970s. After this time, another era began, when energy concerns grew and became more environmental in nature. This revolution would impact all the components of construction: the objectives, standards, design, techniques, actors, system of actors, training, etc.

5.2.2. *Construction methods to improve energy and environmental performance*

In response to the oil shock of 1973, the French authorities favored a policy organized around two axes: the development of the nuclear sector to reduce external dependence (the low-carbon nature of this source was not emphasized at the time), followed by the reduction of energy consumption levels in various areas, including that of buildings. This second part took shape when thermal regulations were first established and were only applied to new buildings. The thermal regulations that followed quickly extended to include old stock too.

The approach to the energy problem that arises from this very restrictive context laid most of the foundations for the technical and socio-technical principles of action that are still relevant today. We can note in particular:

– *Advances in knowledge surrounding building heating* (how does it work?) and the effects on consumption and losses, of heat in particular (Figure 5.1). It is using this knowledge that the first regulation would set out minimum performance standards in terms of insulation (containment, fight against thermal bridges) and ventilation so that humidity can escape.

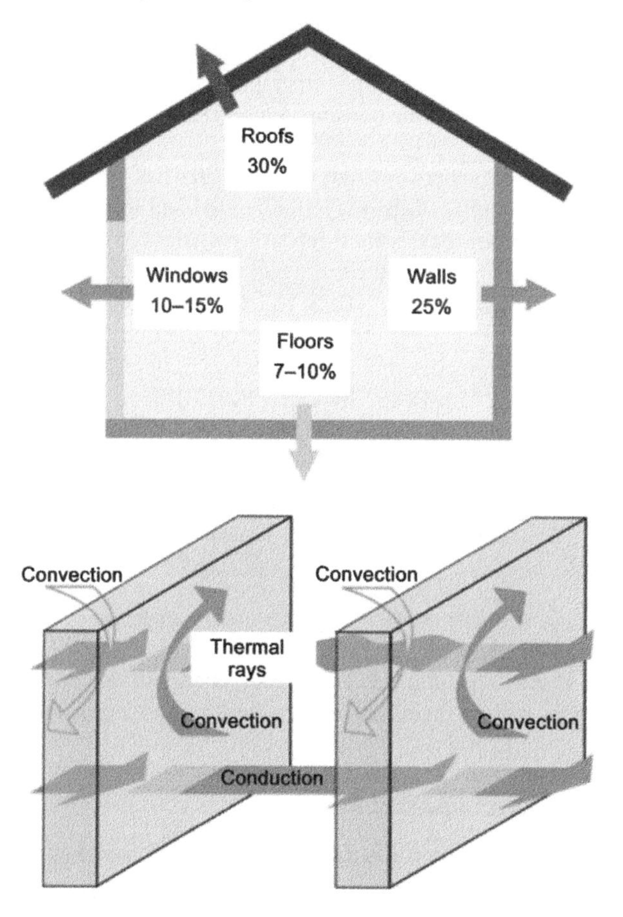

Figure 5.1. *Illustrations of thermal heating in the building[4]. For a color version of this figure, see www.iste.co.uk/robyns/smartusers.zip*

4 Illustration source sites: https://www.travaux-isolation-thermique.fr and https://energieplus-lesite.be.

– *Increasingly efficient technical devices* in terms of performance, operating consumption, efficiency, programming, fine tuning, etc. All technical elements are taken into account: insulation, heating, lighting, ventilation, blackouts.

– *The emergence of thermal calculation tools from IT for modeling consumption parameters*. These make it possible to aim toward theoretical consumption until the reference standards are met. Calculation engines respect the socio-technical nature of consumption. Indeed, for there to be an energy draw, equipment must be occupied and used[5]. Therefore, technical aspects, occupation and use need to be configured. This last aspect is the trickiest because there are as many consumer styles as there are number of households. The solution that is adopted, which is only a compromise, is to use hypotheses of a sociodemographic profile, *standing*, equipment maintenance, time spent within the housing and level of sobriety.

The increase in energy performance requirements has had multiple effects on products, technical solutions, skills, players involved and the relationships between them. On a technical level, this is how repeating operations ended up determining the typical ingredients of a "thrifty" building (insulation, inertia, sealing, optimization of natural inputs, etc.) [BES 15]. Organization and working methods have been particularly impacted by the importance given to the design phase (time spent, budgetary weight, profiles mobilized, level of technicality). This has led us to the point talking about "upstream hypertrophy"[6]. This is mainly mastered by engineers (heat engineers, electricians, acousticians, BIM managers[7]) who are assisted by modeling tools and who always interact with architects. This is a direct effect of the quest for energy and environmental performance, which can be shown by the proliferation of labels.

5.2.3. *Controlling the intensity of energy consumption*

Reduction in energy consumption depends on energy draws, whatever the nature of the building. Controlling them with a view to limiting them is therefore a crucial issue.

5.2.3.1. *The incentive to use less heat*

Because the heating station is the most strategic, as it consumes the most energy, the public authorities targeted it first (from 1974). They proposed a *clear recommendation to limit interior temperatures to 19°C for the entire building stock*: this threshold was said to be adapted so as to necessarily control energy

5 Usage represents 80%–90% of total energy consumption of a building [KOE 07].

6 The authors refer to this as "upstream hypertrophy" [BES 15].

7 BIM, or *Building Information Modeling*.

consumption due to the intensity of the heating (or restriction, in a more accurate sense) and the legitimate expectations of the occupants when it comes to comfort[8].

Also in the same period, new campaigns were launched among populations with a view to *encourage sobriety when it came to the use of energy*. Conceived at the request of the Agency for Energy Savings (AEE) which was created in 1974 (an ancestor of Ademe), the tone of these campaigns has evolved over time: they first used slogans ("The hunt for gaspi", "In France, we don't have oil but we have ideas"), before becoming guilt inducing (19°C is enough), and then educational (+1°C = +7% billed), and finally based on disseminating eco-gestures.

What is sobriety?

The notion of sobriety refers to a search for voluntary moderation, in the production and even more so in the consumption of goods and/or services. The associated virtue relates a reduction in the energy and material resources which are necessary for the provision and use of these goods and/or services.

Historically, the notion has been questioned intellectually in several ways throughout the 20th century. This questioning ranges from "voluntary simplicity", which was defined by the American philosopher Richard Gregg (1888–1974) in 1936, to sometimes radical critical movements, which target the consumer society (Günthers Anders, 1902–1992), through to critiques related to political ecology.

Global recognition of environmental degradation and climatic peril has given sobriety a very post-modern relevance, especially related to an energy variation. Today, it is a question of consuming less and better for one's health, of preserving the environment, of managing one's living environment and of improving well-being for oneself and for future generations. To achieve this, the strategies combine several approaches:

– actions at different scales: society, territories, individuals;

– actions of different natures: laws and rules, infrastructure, practices;

– actions on various themes: energy, mobility, waste, savings, health, food, etc.

This "logic of restraint" which can be applied to energy corresponds to the energy sobriety that is promoted by public authorities and those involved in controlling energy costs at home. But energy sobriety remains a difficult virtue to respect: particularly by households, due to a Western way of life which is still organized around easy access to energy; by people at work, who for the most part do not pay the cost of their consumption.

Box 5.2. *Details about the meaning of sobriety*

8 The origin of this reference temperature was sought, without success [BRI 13].

Of course, awareness campaigns make all the more sense when occupants have control over energy devices. This was and still remains mostly the case in homes as well as in workplaces, with switches, wall thermostats, thermostatic taps and programmers where appropriate.

However, it has always been the imperative of energy performance, supported by the evolution of IT and digital tools, which has modified the logic of the consumption regulation approach. Two major developments can be identified:

– *the need for more detailed regulation, in terms of duration, intensity and programming*, beyond the imperative of complying with the rules already in force (minimum temperature) and a concern to meet needs (health, work);

– *the redistribution of control functions between actors*, alongside a questioning of the settings that can be controlled by the occupants. This scenario has lasted even longer since the control that is closest to the need seemed both logical, easy to implement and in line with what the occupants aspire to [CER 13].

5.2.3.2. *Tools for controlling consumption and potentially power*

The change in requirements and technical means that are available leads to the distribution of various adjustment tools which are technically compatible; home automation may come to mind (particularly in housing), as well as TBM and smart buildings (which exist more in business buildings).

Even though these may not be in the majority today, they are bound to have a bright future for reasons that sit alongside energy and environmental requirements. The experiences that currently exist mean that it is possible to specify the typical circumstances for implementation and the socio-technical problems that emerge as a result. They are summarized below, with a particular focus on the essential critique.

5.2.3.2.1. Home automation: a tool for energy and mental comfort

The promotion of home automation is closely linked to the domestic distribution of control terminals (computers, tablets, smartphones). The commercial success of the tool is due to the fact that it speaks to very contemporary concerns: increased freedom of movement, decreased mental load because of programming, more effective management of energy and finances, reduction of the environmental impact, monitoring of remote housing, etc.

The real risk of home automation relates to the rebound effect. This notion evokes the emergence of an increase in consumption which is induced by a capacity for more effective management. Clearly, a household may be encouraged to consume more, propelled by the feeling of energy uses being better rationalized.

Additional to this are the paradoxes that are associated with home automation. For example, it is in and of itself a source of consumption. Automation can be used without actually being at home (including heating automations). Certainly, home automation is synonymous with practicality and comfort, much more than energy and financial savings may be. Table 5.1 contextualizes home automation by presenting how it is applied, the circumstances and contributions that favor its application and points of weakness.

Typical field of application	– Households of a higher socioeconomic status – Motivated by the environment, controlling costs, less often by technology alone
Favorable circumstances and contributions	– Growing trivialization of connected objects – Source of social distinction – Social body that is increasingly concerned about the environment – Potential coverage of all energy and security posts
Points of weakness	– Feeling of control which facilitates a possible rebound effect – Inflation of a social model based on possession, with new forms of consumption – Fear of hacking (loss of control and data theft)

Table 5.1. *Contextualizing home automation*

5.2.3.2.2. Technical management of a building: support for optimization and dialogue between technicians and occupants

TBM facilitates supervision, maintenance and needs-based adjustments. Energy draws can be better identified (function, devices, intensity, duration, location) and therefore better understood. It is a valuable computer tool which helps professionals who are responsible for the building. Table 5.2 contextualizes the technical management of the building.

Typical field of application	– New or renovated activity buildings – Mainly tertiary buildings with staff
Favorable circumstances and contributions	– Adjustments for comfort, potentially in a nuanced way – Little technical mastery (possibility of autonomy) – Easy maintenance (alarm, diagnosis, rapid intervention) – Ideal for large and/or complex buildings (technical devices, regular changes in needs, etc.)
Points of weakness	– Loads for management and valuations (data formatting, consumption analysis, knowledge of needs) – General risk of underutilization (few functions monitored, few data analyses undertaken)

Table 5.2. *Contextualizing the technical management of the building*

From a socio-technical point of view, the TBM is innovative in that it harbors the potential for dialogue between technicians and occupants. Indeed, producing data facilitates various optimizations, all with the starting condition of paying attention to the occupants and their needs. Typically, this is about:

– adjusting draws according to actual usage;

– adjusting consumption intensity according to the level of satisfaction expressed;

– undertaking awareness-raising, correction and/or optimization actions according to the anomalies identified according to the size of the draws.

To properly implement the potential for optimization, we first need sufficient resources to enhance the data produced (formatting, analysis, optimization), before collecting a minimum amount of data about needs (understanding of the site and how it is perceived by occupants). However, taking an interest in these matters before directly addressing the occupants is not a natural action taken by professional technicians, who are often not made aware of this aspect during their initial training, or which does not constitute a priority of their practice. Additional to this unfavorable context is the fear of what an initial expression of dissatisfaction on the part of the occupants can represent if the expected improvements are disappointing. All of these circumstances serve to explain why the dominant management model of a building with GTB remains splitting the concerns of the technicians from those of the occupants, who continue to use energy without information, guardrails or real support.

The result of a survey conducted on a university site with high-performance buildings in 2015 illustrates this. It is marked by the perceived importance of the technical culture as well as the limited openness from the occupant's point of view:

> [For] the technical agents [...] who are responsible for the proper functioning of the installations [...], the innovative character of the building has had very little impact on their professional practices. The objective of their mission seems to prevail over the means of achieving it: it is expected from their work [...] that the installations operate without interruption for the technicians [SUB 15].

5.2.3.2.3. Smart buildings

Smart buildings are characterized by technical-energy management which takes advantage of the possibilities of IT and digital technology. They differ from TBM in that they allow one to optimally integrate renewable energy (REn) and they have superior capacities to exploit building operating data in order to adjust consumption to meet needs.

From a socio-technical point of view, the central question associated with smart buildings is that of the position granted to the occupants. Indeed, the inherent capabilities of the tool mean that the entire management of the building can be attributed to digital mechanisms (data processing, algorithms, predictability, optimization, etc.), rather than to associated occupants whose varying and partly unpredictable behavior constitute the major uncertainty of energy performance. Table 5.3 contextualizes the smart building.

Typical field of application	– New or renovated activity buildings – Ideal for local energy production
Favorable circumstances and contributions	– Straightforward greening of the electrical network – Facilitated service offer (supply/demand management) – Perspective of added values (energy, comfort of services)
Point of weakness	– Risk of magical technological thinking – Logic which is not very compatible with delegating settings to users – Risk of rebound effect by consumption of internal installations

Table 5.3. *Contextualizing a smart building*

Hindsight and feedback are not yet sufficient to be able to claim that digital mechanisms embedded in a smart building are capable of gaining more control of energy consumption without involving the occupants, as well as more self-consumption if energy is produced locally, a better quality of use, more daily satisfaction, etc.

On the other hand, it is certain that putting the occupants to one side in the management process of a building, which seems like an efficient option, proves to be counter-intuitive. We know that the higher the performance of the building, the greater the impact it has on users:

– Through increased energy draws, due to the performance of the "box" which operates from a minimum level of consumption. The reminder mentioned at the beginning of the chapter regarding an occupation that is synonymous with consumption takes on an even stronger meaning in the context of the claimed performance.

– Through the instructions for use which are to be followed. For example, opening windows inconsiderately disrupts the balance of heating and ventilation, or the combination of insulation and sun requires one to use blackout curtains at the risk of overheating.

These two illustrations will be supplemented later by other findings from sociological research. Yet, they already attest to the difficulty or even the impossibility of respecting energy requirements without giving a role to the occupants.

5.3. Determinants of energy use in the world of work

Regardless of the building, energy consumption depends on the interaction between a construction with varying levels of "efficiency" in terms of energy, and an occupancy with varying levels of "sobriety", that is to say, one is concerned how much they are consuming. Efficiency is easier to control, since the effects produced by the combination of construction options can be modeled according to determined constraints (consumption, temperature, operating ranges, etc.). It also comes from the best practice parameter. The latter are in fact due to human factors, which by nature cannot be easily modeled, but whose awareness-raising actions have attempted to frame in order to make them comply with technical requirements.

It is accepted today that the most common framing strategies (technical information, instructions for use, reminder of the climate emergency, etc.) struggle to obtain the expected practices. Or to put it another way, the "cooperation" of occupants to achieve the objective of reducing consumption. The gap between expected and actual uses is undoubtedly the major challenge posed to any socio-technical situation which targets a given level of consumption. To better understand the driving forces behind this, it is possible to put forward the main explanatory factors identified through research work:

– the vocation of the company and the predictability of its work;

– energy consumption outside performance criteria;

– the search for comfortable working conditions;

– sensitivity to environmental issues;

– differences in energy sobriety at work.

Reviewing the main production factors of energy practices will make it possible to contradict the belief about undisciplined, resistant or indifferent occupants when it comes to reducing energy consumption. Practices that are deemed unconventional are not based on a difficult relationship with rules or standards.

5.3.1. *Determinants linked to business activity*

5.3.1.1. *The vocation of the company and the unpredictability of work*

An organization (company, administration, association) is first and foremost a project that requires a material and functional framework (building, equipment, work organization, working conditions that comply with legislation, etc.). Once this framework has been produced, work begins and starts to unveil what is called the "activity system". This corresponds to how the company functions, from a material and human point of view, and how it has been planned out.

Leveling out the activity system helps one to identify a first set of determinants related to the socio-technical energy problem within the organization. First, the equipment is taken into account (considering whether it is numerous or scant in number, consumer-facing, used, or shared), followed by when activity is undertaken, and the temperatures associated with the vocation. For instance:

– a medical analysis laboratory has few staff, but many machines, some of which operate at night, have no temperature requirement;

– a university which welcomes the public is confronted with various imperatives to adjust temperature according to how the spaces are used (static positions in offices, high temperatures and compromised air quality in classrooms).

Components	Contents	Impacts/energy demand
Vocation	Open to the public? Storing food? Preservation of samples? Hours of operation?	Need for heating? Need for power? Continual power supply? Secure supply? Need for lighting? Security contraints? Regulatory requirements?
Stage	Number? Diversity? Position? Physical activity?	Attendance time? Stability? Permeability to instructions?
Work situation	Exterior/interior? Day/night? Mobile/static?	Heat and light contributions? Working conditions? Regulatory requirements?
Regularity of activities	Operation hours? Seasonality? Regularity?	Chronomanagement? Programming? Anticipation of needs?

Table 5.4. *Components of an activity system*

It is also wise to *anticipate attendance times* for the programming of energy devices (also called "chronomanagement"). For example, teaching sites operate with lesson schedules that can be used as a reference to modulate heating and ventilation in the rooms in question. Office heating can also be programmed based on regular schedules. Of course, advance programming has its limits, with attendance times that differ from those announced, hence the interest of coupling chronomanagement with presence detectors. Table 5.4 summarizes the components of an activity system.

There are other sources of delay that are more difficult to overcome. This is typically the case with the layout of workplaces configured as shared offices (of usually two to four). If we consider the optimal scenario, heating will be designed so that it takes into account the heat contributions that are associated with the presence within the building (body heat and IT in operation). Having said this, it is enough for people to be absent for a certain time for the ambient temperature to drop because of an undersized heating system coming from thermal hypotheses.

These explanations help one to understand why the sophistication and performance of a building do nothing to protect against the risk of thermal discomfort. The problem is that the feeling of discomfort always provokes a sense of dissatisfaction, protestations, then a search for spontaneous solutions which often impact the thermal diagram of the building as well as its energy consumption (opening of windows, auxiliary heating, etc.).

5.3.1.2. *Energy consumption outside performance criteria*

Energy issues are set aside in most organizations, as if it is a blind spot detached from the reality of work. Of course, incentives for sobriety at work are developing. Similarly, organizations are committing to an ISO 14001 labeling process designed to reduce their carbon impact. Apart from these two trends, daily work is marked by an absence of demand or control over the energy issue. Pertinent issues are more related to efficiency and success indicators (work speed, waste, blockers, customer satisfaction, cost control, etc.).

Real working conditions related to energy are the product of interdependencies between regulatory frameworks, employee concerns and the need for work to get done. The question of energy is thus relegated to the background, whether unconsciously or due to fatalism.

5.3.2. *Sociological determinants of energy practices*

Sociological research attests to the contextualized and individual nature of ways of thinking about and acting using energy in the world of work. *It is always a question of "feeling", but according to different registers*:

– *the primacy of thermal comfort, above all in cold weather*: having enough heat is a primary need, a factor of good health, of comfort at work and incidentally of efficiency. The cold felt depends on the measurable temperature of the space that is occupied and the nature of the work, as movement and effort increasing thermal resistance. In the same building, there are always temperature differences that are due to sunlight, quality of the windows, the possible undersizing of the heat emitters, the presence of a heat-emitting machine, etc.;

– *sensitivity to environmental issues at work*: this determines the permanence and extent of the sobriety deployed on a daily basis in the use and/or interaction with energy-consuming devices. It also influences the practices that people will strive to mobilize to counter thermal discomfort in particular. Some of these practices will add to energy consumption, as is typically the case with auxiliary electric heaters;

– *the relationships experienced in the context of work*, above all with the people in charge of a management or supervisory function, but also with colleagues and peers. They sometimes exert a strong influence over what is called "organizational citizenship behavior directed towards the environment" [LAB 15].

5.3.2.1. *The primacy of thermal comfort at work*

To be experienced in a satisfactory way, a work situation requires a minimum of good conditions. Some are related to how the work itself is carried out (adapted equipment that works effectively), others relate to working conditions in the broad sense (space, lighting, temperature, noise, smell, etc.). All benefit from a regulatory framework[9]. Designers, maintenance personnel, company managers and employees thus benefit from useful reference technical standards to assess what needs to be done, checked or claimed. But in the range of working conditions, an appropriate ambient temperature which provides a feeling of comfort and ability to work is the priority concern of the occupants of a building.

In any business building, thermal sensitivity is always the strongest because it is at the crossroads of different dimensions: physiology, health, productivity, symbolism. This explains why, placed in a work context which is experienced thermally as uncomfortable in winter or summer, occupants can be very active and imaginative when it comes to correcting a situation that they experience poorly.

Figure 5.2 presents the energy consumption levels by user station in a building.

---

9 For example, in the Labor Code, article R4223-13 for thermal comfort, decree No. 2008-244 for lighting.

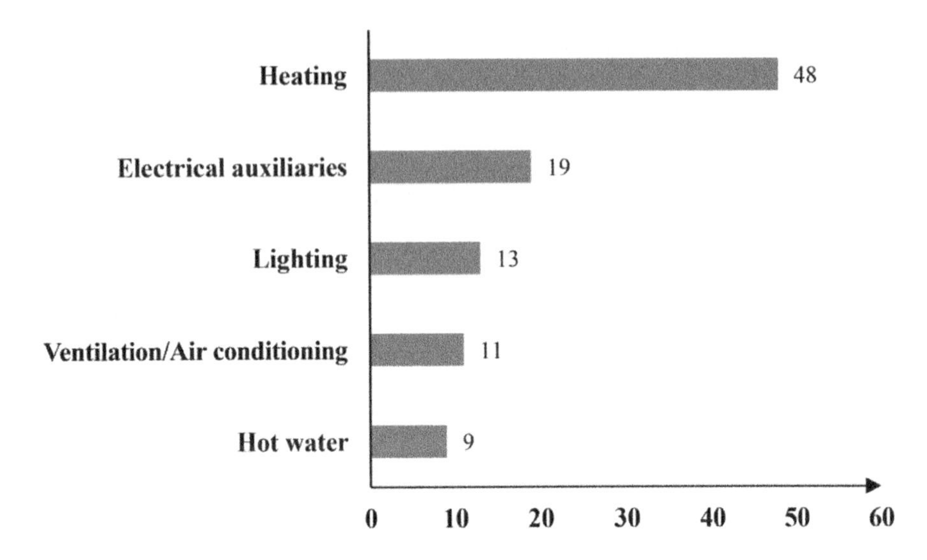

Figure 5.2. *Percentage breakdown of energy consumption by station of use in the tertiary sector in 2013 (source: Schneider Electric)*

5.3.2.2. *The solution for manufacturing thermal comfort*

In a situation of thermal discomfort, typically in a cold period, the simplest solution is to use an auxiliary heater bought out of your own pocket. This also allows you to be ready for anything in the event of a change of office[10].

The use of notoriously high-consumption electrical appliances is a widespread practice, not least because it is easy and efficient. Operating rules of organizations are in fact ill suited to the production of a specific and localized solution: in winter for example, it is difficult to change a radiator that is too small or a damaged window without going through a large-scale call for tenders, which takes time and requires a suitable budget. In the same way, in summer the isolated installation of a blackout curtains of decent size and filtering is not a popular option with technical services[11]. In this context, the acquisition of devices is no longer simply a problem from the point of view of the consumption it causes, but a symptom of

10 Following a survey about energy practices on the campus of the Université Catholique de Lille, it is estimated that at least 45% of respondents have an office equipped with an auxiliary electric heater. Many can benefit from this very device.

11 The author recalls the difficulty experienced by those working for a French minister to buy a simple coffee machine. After the exchange of several official internal letters, a line of credit was created with an oversized budget…

organizational maladjustment in the face of technically insignificant difficulties, which exert a noticeable cost on the business.

Sometimes maintenance distributes technical aids (heaters in winter, fans in summer), but rather as a last resort to calm the occupants of a cold area. Providing auxiliary heating for a particular office is treated as a one-off event so as not to set a precedent that could cause requests and bills to explode[12]. Finally, we see that organizations are locked into unsuitable treatment methods to resolve isolated thermal discomfort; everyone is forced to "tinker" with a desired effective solution, in winter as well as in summer.

5.3.2.3. *Extremely varied thermal solutions*

The work context is no different from the same situation at home, for example, in cold weather: "What can I do to be warmer in my office? Or over the whole day? But definitely in the morning when you first arrive?" The difference is that developing the solution is done by combining the desired thermal result with the organizational context, which does not allow everything and anything to be carried out. Indeed, there are fewer objects or tools in work than at home, and circumstantial constraints may exist: avoid degrading the surroundings, "build" without being noticed and enjoy the fruits of one's ingenuity discretely.

"Tinkering" to increase comfort in a given situation has been done for as long as a habitat for the human species has existed. The notion has been taken up in many research works, and recently reformulated by the term "construction" to mean what happens daily within a spot of housing.

Table 5.5 lists the most common thermal "constructions", including the straightforward use of an electrical appliance, auxiliary heater or fan. The solutions are certainly artisanal, but they are imprinted with good thermal sense. It is a question of heating, partitioning, ventilating, ventilating or cooling.

This table shows the most common tricks used in the workplace, and we can obviously see how practices are transferred from the domestic world (use of hot drinks, candles, etc.). The use of electrical devices was placed at the top for both seasons because they are the most spontaneous, efficient and energy-consuming actions.

The diversity of solutions implemented between people refers to the inequality of "energy intelligence" when it comes to comparable thermal problems. But this in no

12 The notion of the "rebound effect" is valid when there is new consumption which is carried out under the pretext of an efficient building. The case described concerns the simple implementation of a consumer solution.

way detracts from the organizational interest inherent in the solutions, quite the contrary. Apart from the use of electrical appliances, *"building" makes it possible to break with the almost mechanical relationship between comfort and energy consumption*. Activity is another major source of savings that results from practices with eco-gestures that can be applied to the office[13].

Problems to solve	Recurrent workplace solutions
To be warmer	– Auxiliary heating – Join the windows with tape – Close the curtains during the day – Layer up on clothing – Cover yourself with a blanket, which is quick to take on and off – Use filament lights that heat up – Install a bellhop to close the door properly – Light candles – Consume hot drinks – Block the ventilation in the ceiling or the windows – Place a cool object on a clamped thermostat – Arrange your space to avoid having your back to the window – Make use of comfortable spaces or equipment
To be cooler	– Install a ventilator – Decompartmentalize by blocking the door – Add plants to your space – Reduce indirect heat gains – Move to a cool place – Make sure a window is open at night – Cover the windows

Table 5.5. *Tools to improve thermal comfort*

5.3.2.4. *Sensitivity to environmental issues*

Opinion polls show that society is now aware of the dangers associated with climate change and GHG emissions[14]. This comes following several decades during which a contemporary way of life has gradually ignored, made space for and eventually accepted the need to take environmental externalities into account. The evolution mentioned here can be found in attitudes and behaviors related to various means of consumption (food, transportation, leisure, etc.). All that relates to energy

13 Essentially, this involves turning off computer equipment and printers at night, turning off lights in unoccupied spaces, using stairs and avoiding opening windows during heating periods.

14 Excluding Covid, protecting the environment was the number one concern of the French in August 2021 (41%). Source: IPSOS-Sopra Steria survey for *Le Monde*, the Jean Jaurès Foundation, the Montaigne Institute and Cevipof.

is also concerned, since *energy sobriety (the idea of paying attention to one's energy draws in order to avoid or limit them) has become a reality for the majority of people*. Does this mean that the level of sobriety declared can be verified in the context of work? Where will we see positive effects on consumption and compliance with instructions for use, including within smart buildings? Not quite.

The context of energy consumption at work is indeed different from the domestic context, so much so that sobriety declared in these spaces is in decline (10% loss according to surveys), and it is undoubtedly less constant according to the times and the places of use. This gap can be related to three purely organizational parameters:

– Energy bills are paid by companies and not by their staff. The price factor already has little impact when it comes to housing due to imperative needs and a lack of alternatives [CRE 13]. This is even less so in a work environment.

– The occupants are not at work to save money, and they are never or rarely asked to do so under the pretext of controlling costs. The expectation is above all to benefit from a building that is adapted to the requirements of their work [CER 13].

– Paying attention to your consumption is a mental load additional to that exerted by work. It requires a triple effort linked to design (how to act while consuming as little as possible), consistency (to maintain a sense of comfort and energy vigilance) and assimilation until it becomes routine. For all of these reasons, chosen or imposed sobriety is at odds with a sense of ease.

5.3.2.5. *The enigma of energy sobriety*

The existence and intensity of sobriety at work is not only a matter of individual will. It entails a combination of psycho-sociological factors that reinforce each other until they orient and anchor behavior in sobriety. These factors (or parameters) are distinguished below. Figure 5.3 offers a representation of the energy sobriety parameters.

– *First factor*: a key component to adopt a sober practice is *adhering to the idea of virtuous energy sobriety from an environmental perspective*. This becomes even more important when evidence has been gathered from various sources (documentaries, testimonies, reports, etc.), which can reinforce the effects of a related personal story. Once one has got into the practice of sobriety, it becomes rational and achievable despite the psychological and/or physical efforts that it may induce.

– *Second provision*: a key component of sobriety can be articulated related to a *feeling of justice experienced at work*, which is called "organizational justice". This feeling can be easily measured because it is based on three criteria:

- procedures deemed to be clear and applied in the same way for everyone, that is, “procedural justice”;

- distribution of the company’s resources (premises, materials, finances, etc.) which are deemed to be coherent because they are proportionate to the contribution of each person through their work, that is, “distributive justice”;

- superior/subordinate relationships judged to be characterized by honesty, respect, benevolence and transparency, that is, “interactional justice”.

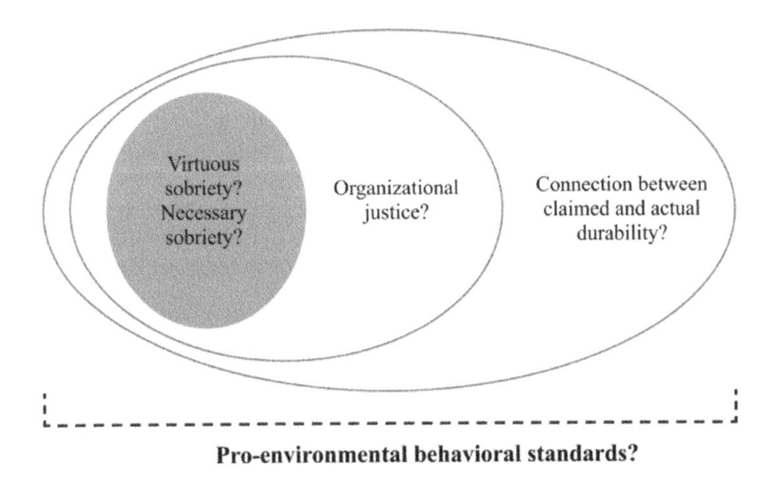

Figure 5.3. *Parameters of energy sobriety*

Research shows that the higher the sense of justice, the more likely staff are to adhere to the pro-environmental practices promoted by the company [LAB 15].

– *Third factor*: this depends on whether the sustainable functioning that is demanded by the organization and what is proven when it actually functions are consistent. If coherence is obvious, occupants are more inclined to adopt sober energetic behaviors. If not, the contradiction between what is desired and what is observed causes a misunderstanding, followed by a psychological discomfort on the part of the occupants who are either sober or ready to become so, and eventually a downward adjustment of behavioral sobriety among those who experience this contradiction most poignantly.

– *Fourth provision*: this concerns *the normative influence exerted by the occupants of a building on each other*. Indeed, people at work live in different embedded communities (the company, the department, the team, the shared office, the job, etc.). Each has a set of social norms that will influence practices, most specifically when it comes to imitation (behaving like others) or development (doing what is expected). Of course, *social interactions at work also produce energetic behavioral norms that will or will not result in sobriety*. The preferred practice at the individual level results from a subtle appreciation of what there is interest in doing when it comes to the company, one's colleagues, one's peers (at the risk of formal sanctions and/or relational ones), and what one's intimate values require on the energetic level.

5.3.3. *Modeling the belief-behavior relationship*

5.3.3.1. *Energy practices and sobriety*

The number and diversity of influences that are exerted on human behavior make their understanding always very complex. The energy field is no exception to this rule. In order to try to approach the enigma of behavioral dynamics in a built environment in spite of the complexity, and more specifically in the tertiary sector, we have prioritized a few aspects that seem essential to us. These are given below:

– *The imperatives of energy consumption which are controlled by the company's activity system*, with particular attention on the predictability of work which relates to the constraints of energy supply (availability, intensity) and the degree to which the constraints that come from work (rhythm, deadline, quality) and the desire for sobriety of each staff member can be reconciled. Faced with these consumption imperatives by activity, there are three consumption levers: the building, behaviors and the organization of work.

– *The importance of thermal comfort for the occupants, depending on the building*. When comfort is absent, this most often leads to "construction" strategies, which have varying levels of ease to implement, sophistication, efficiency, duplicability, and energy-consumption.

– *The constancy of sobriety in energy practices*. Each occupant determines this according to the organizational context that is experienced (work imperatives, adherence to the company's expectations, influence of the relational system) and according to their ability to accept to act with less sobriety if the circumstances are right. Value systems play a central role here: the stronger one's attachment to the environment, to health or to intergenerational transmission, the less one will agree to diverge from the principle of sobriety, whatever the circumstances and the consequences.

The relationship identified above between thoughts (beliefs, representations, attitudes) and behaviors (sobriety, "construction", constancy) makes it possible to *formulate a case model around two axes*:

– *axis 1 (horizontal) conveys beliefs* regarding the climate emergency and the need to act. The power of adhesion directs practices toward sobriety;

– *axis 2 (vertical) conveys behaviors* that are potentially sober, constant and fully aligned with beliefs.

The intersection of the axes leads to one to construct a typology which is made up of four distinct tendencies (Figure 5.4). They each bring together individuals with similar beliefs and behaviors. The attachment to the current climate (an alarming situation?) and energy (interest in sobriety, a more sustainable way of life?) facilitates attachment to one of the trends or another:

– *First profile, the ultra-sober (+/+)*: they are the most convinced of the urgency and usefulness of acting, at all times, including on a small scale. Their conviction sharpens their desire to always work with concern for absolute sobriety, which accentuates their ability to find the appropriate solutions, by mobilizing their energy intelligence to take action and without neglecting their thermal comfort. For them, there is little or no paradox between their pro-environmental values, their desire for sobriety and their actions.

– *Second profile, sober legitimists (–/+)*: the intensity of their belief in the climate emergency is not very strong, and in the end, it does not have an impact. Their attachment to the world of work and its rules is enough to make them follow the sobriety incentives issued by their organization. They respect what they are encouraged to do, without having to demonstrate energy intelligence, hence the need for assistance through socio-technical advice.

– *Third profile, workers first and foremost (+/–)*: their environmental convictions are stronger than those of the "sober legitimists", but they do not really take action. They do not work to save energy. Rather, their primary motivation is related to what they do and what they get from it (remuneration, esteem, pleasure, etc.). Motivated in their work, they will willingly reduce their alertness to energy if it becomes detrimental to efficiency and comfort. Their sobriety is therefore variable, while it can be more constant at home.

– *Fourth profile, indifferent, in the background (–/–)*: at best, the climate emergency is not their problem, at worst, they do not believe in it. They therefore choose not to take sides by working as they have always done, without any particular concern for sobriety. Their presence in the company is very instrumental, with their primary motivation being to earn a living there.

Modeling four typical profiles suggests a graduation of sobriety in energy practices. On the basis of available data, a distribution hypothesis can be put forward:

– In the population, environmental concerns dominate, which leads one to think that sober workers are in the majority (from 60% to 80%). Among them, the “ultrasober” are in the minority because of the intensity of their beliefs and above all their commitment to an individual trajectory with the least carbon content possible (from 10 to 30%), at home and at work.

– Occupants with little or no sobriety are in the minority (from 10% to 30%), with a large proportion of people whose sobriety is modulated according to priorities, as is the case for “workers first and foremost”. Those who are most indifferent to climate data and sobriety have become very few (less than 10%). They represent a hard core that is difficult to develop.

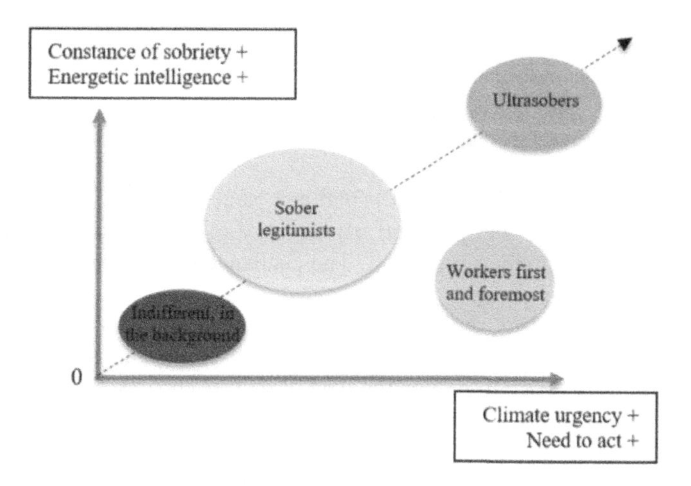

Figure 5.4. *Typology of intensities of sobriety*

5.3.3.2. *The impact of a building in the form of a smart building*

The building has hardly been discussed so far. This is a deliberate choice as our objective has been to understand energy practices at work, to give priority to the invariants that make up the organizational context (activity, management, relationships), the importance of thermal comfort in the eyes of the occupants and finally the influence of beliefs when it comes to adopting a variably stable and strong sobriety. As this has been achieved, it is important to highlight the significant role the built environment plays in consumption; it is primordial. This is because if there is no consumption without occupation, it is indeed the building (with its construction and socio-technical amenities) that explains most of the magnitude of the energy draws in a given climatic context [LÉV 14].

If smart buildings boast a significant energy promise, their socio-technical conditions do not seem to be clear cut. There is still uncertainty among designers when it comes to the value of putting occupants in the performance feedback loop. This unknown poses questions that must be addressed in order to estimate the potential, assets, risks or uncertainties of these buildings based on the combination of IT and digital forces.

5.4. High-performance buildings abused by uses

The contemporary debate about the contribution of occupants to the energy performance and quality of use of buildings is predominantly technical and not sociological in nature. It refers to the belief in achieving the best performance by almost exclusively mobilizing the characteristics and functionalities contained within a building. The question concerning how respective contributions between technology and humans around energy are balanced is not a new one.

It is even possible to affirm that it is part and parcel with the fact that apart from sunrays, the production, transport and maintenance of energy forces over time are always a matter of technique and effort. The challenge of lightening the load borne by humans is therefore not new.

The impact of this age-old concern has now led buildings to be equipped with elements that are designed to lighten the load of their occupants[15] as best as possible [FOU 66].

And since there is a wide variety of solutions to achieve this reduction, it is useful to demonstrate them based on recent achievements. For this, we have chosen to rely on two buildings present on the campus of the Université Catholique de Lille, which share the same university vocation, but whose socio-technical philosophies are radically different.

5.4.1. *The philosophy and equipment of the building*

5.4.1.1. *The Rizomm demonstrator building*

The Rizomm is a teaching and research building located on the campus of the Université Catholique de Lille (Figure 5.5). Built in several stages since 1951, its

15 The concept of “historical viscosity” designates the continuous process of transformation of the social in the context of the moment.

current total area is 7,700 m² spread over five floors and assigned to four major functions with a particular purpose: offices, classrooms and amphitheaters, meeting and work rooms, and common areas for relaxation and hygiene. The building was completely renovated between 2016 and 2018 according to set specifications allowing it to be recognized as a "demonstration building" by Mission rev3, the coordination unit for the deployment of the Third Industrial Revolution in the Hauts-de-France region (see section 3.4.4.2).

Figure 5.5. *The Rizomm building at the Institut Catholique de Lille*

5.4.1.1.1. Performance biases including users

The performances that were targeted when renovating the Rizomm cover two major area. The first relates to energy, with the challenge of reducing consumption as much as possible and maximizing the self-consumption of solar energy that is produced on the roof of the building, in order to have as low carbon operation as possible. The second concerns including users in the fine adjustments of heating and office lighting. The main technical and socio-technical aspects of the building are detailed in Table 5.6.

The Rizomm renovation incorporates major innovations for the campus and for energy management of the building. As such, it is a real smart building, which is equipped with trigger sensors (lights, heating, toilet water tanks), a TBM that regulates the ranges and intensities of energy draws (i.e. chrono-management) and a decentralized supervision system for electrical self-consumption.

	Technical details
Energetic performance	– Control of solar gain – High efficiency LED lights – Windows fitted with heat stoppers
Self-consumable energy production	– 1200 m^2 of solar panels – Energy recovery lifts
Optimization of consumption by TBM	– Detailed adjustment in each space (heating, light, ventilation) according to regulatory, contractual or management references – Increase in the set temperature by presence sensor during the day, from 19 to 21°C – Triggering of lighting by presence sensors, varying according to light intensity – Presence sensors for resetting the water circuit in sanitary areas – Timing of heating based on office and course schedules – Monitoring of consumption and temperatures according to the actions of occupants
Interactions with users	– Variable heating in offices and classrooms (+/–2°C) – Presence of a facility manager for monitoring equipment (operation, performance), TBM and supporting occupants (technical explanations, responding to needs) – Screens that summarize the energy consumption and solar production of the building

Table 5.6. *Detail about the socio-technical devices in the Rizomm building*

5.4.1.1.2. The regulation of interior temperatures by the occupants

The Rizomm philosophy also integrates the users as agents for regulation in two ways:

– for heating, with the possibility of making a set temperature of 19°C vary by 2°C;

– control of solar gains, with a simple switch that allows for a blackout blind to be raised or lowered as needed. Permanently, the presence sensor adjusts the light intensity to 500 regulatory lux.

5.4.1.1.3. Recruiting a facility manager

The greatest innovation associated with the Rizomm is undoubtedly the creation of a facility manager position, the first of its kind on campus. The function includes regular monitoring that the installations are functioning properly, monitoring the GTB to optimally adjust the settings to needs and expectations, and finally being in contact with the users of the building, who must understand and appropriate the socio-technical devices. This organization of building monitoring breaks with more traditional maintenance methods. In other words, it is more sensitive to malfunctions between two equipment renewals.

5.4.1.2. *The building of Higher Engineering Studies (HEI of Junia)*

The construction of the school building was completed in 1885. Most of its 13,700 m^2 surface area is divided between classrooms, amphitheaters, work rooms and offices. Since engineers need to be trained in the building, it also houses a variety of technical equipment, which consumes a lot of energy and incidentally releases heat. Like the Rizomm, the HEI building (Figure 5.6) became an effective smart building during 2019–2020, after major transformations including complete thermal renovation completed in 2014, and the provision of various equipment which is still being adjusted in some instances (TBM, sensors, electrical supervision center, etc.). The stakes for the school are high. The know-how offered within the engineering courses needs to be showcased, including the building, IT and digital, electrical engineering, and more recently a sector for smart grids and smart cities. The building is therefore also presented as a demonstrator of how solutions are implemented which offer a level of control of energy consumption and high quality of use for the occupants. Implementation is expected to reduce consumption and high quality of use by around 10% (see section 3.4.4.2).

The technophile center of this site highlights that the building's renovation was executed with a preference for automatic features. This means that this particular site typifies technophile thinking when it is applied to a smart building. The occupant is expected to do nothing except accept being subjected to trigger sensors and/or regulators, as well as report any malfunction to technical services. The remarks made by an engineer from a company involved in the HEI building project clearly set out the distribution of roles between technical devices which are in charge of everything and the occupants (presented here as smart users) who must be comfortable to take advantage of the technology utilized: "A smart user is a user who does not need to take care of anything. When they enter a room, the technology adapts to them, to their profile. For example, when a teacher enters a class with their students, the luminosity and the ventilation adapt. The teacher can thus concentrate on their tasks without worrying about their environment".

Figure 5.6. *The HEI of Junia*

5.4.1.2.1. Performance through automation and programming

In the HEI building, the "social part" of the devices can be summed up mainly in two parts. The following description emphasizes the absence of direct intervention by the occupants, except by their mere entry into and presence in the spaces which are thus equipped:

– in the upstream part of occupations, through programming that is aligned with the legal obligations of working conditions and settings that are intended for the triggering, regulation and shutdown of devices according to what the various sensor networks detect (presence, human density, opening windows);

– during occupations, by the activation of the sensors which start, modulate and stop what has been planned by the programming.

Table 5.7 details the socio-technical devices in the HEI.

	Technical details
Energetic performance	– Window opening sensors to turn off the heating – Triggering of lighting by sensors, before turning off after 15 min. Switches always being accessible – Room management programmed according to equipment and heating mode (radiant, ventilation heating) – Adapting temperatures of the heating networks according to the temperature outside – Adjusting natural lighting in the reception hall
Air quality	– Dual flow air handling units – Planning the starting up by chronomanagement – Intensity adjusted by sensors
Optimization of consumption by TBM	– Detailed adjustment in each space (heating, lighting, ventilation) according to regulatory, contractual or management references – Chronomanagement of heating based on the schedules for using the rooms and the number of people expected. Verification of an effective presence by detectors – Triggering of lighting by sensors
Interactions with users	– No decentralized adjustment, but presence sensors to trigger (heating, lighting) or adapt the intensity of the installations (air quality)

Table 5.7. *Detail of the socio-technical devices of the HEI building*

5.4.2. *Feedback from uses in smart buildings*

This feedback will be purposefully limited to aspects of use, without studying the evolution of consumption. It is indeed always very difficult to explain the variations in consumption in university buildings. Their occupation and the pedagogical methods involved in consumption are constantly changing. A meticulous analysis is a job in itself that still needs to be done. Furthermore, it takes an average of 3–5 years for such massive buildings that claim to be efficient to be fully equipped as planned and for them to be stabilized on all technical levels. But before fully addressing the two usage returns, two important remarks must be made.

First, we want to specify that this exercise has been approached from a sociological perspective, focusing mainly on the occupants. This means that the challenge becomes how to assess how users interact with their socio-technical environment and what this environment produces in terms of quality of use. The approach aligns with the description made by an energy sociologist who was mobilized on a Belgian efficiency operation: the challenge “is not to question energy ambitions, but to confront the promises made about the buildings’ performance within the context of the user experience, in order to move towards better mastery of these technologies" [MÉT 17]. Within this logic of critically analyzing a given use case, the smart nature of buildings or the equipment becomes all the less important compared to the devices, which often have sensors and operate according to the settings configured before use. Thus, in this case, feedback is based on testimonials from occupants who come across devices with their current settings:

– The reference period of the Rizomm goes from 2018 to the beginning of 2020. The equipment operation was at that point close to the final objective. In this way, we can say that the context of use was that of a smart building.

– The usage survey at HEI took place at the beginning of 2020, that is, in a building with equipment that was not fully operational. For this reason, the results should not be considered as issued from a building that can be fully qualified as a smart building.

Second, it should be noted that the feedback around usage questions the quality of device appropriation more than their acceptability. The notion of appropriation is less well known although it nevertheless corresponds to most socio-technical situations. For example, when there is a device, whose functioning must be understood in order to use it (with or without assistance as per a user manual) and in order to achieve the desired result, whether or not technical prescriptions have been respected. The fact that this process of "discovery-use-result" does not always allow the user to achieve their objective makes it clear that appropriability cannot be reduced to their capacities alone, but also it depends on how likely it is that the technique can become more aligned with what the user is able to understand and/or undertake. The path that technology must take in order to be accessible to uses corresponds to what sociology calls appropriability and the philosophy of science "the socialization of techniques".

Acceptability is a concept that is much more present in the ambient discourse. It is often put forward when it comes to arousing the social support of a population (e.g., for the installation of wind turbines) or of a target audience when launching a product. Acceptability is very different from appropriability in that its dynamic is more collective and it relates to socio-technical issues where the margins of adjustment are less related to the object itself than to exogenous arguments (jobs, finances, nuisances, etc.).

When analyzing the conditions of use that we want to undertake, the reference criterion is more that of appropriation than that of acceptability: devices have been imagined, designed and installed by engineers, with varying levels of intended mobilization of the social sciences. Once these devices are in place, users must understand how they work, and make the most of them according to what they want to do. The observation concerns as much the capacity of the occupants to exist with their socio-technical environment as the flexibility of use which is offered by the devices within the buildings. The exercise should shed light on understanding what a smart user can be in terms of attitudes and practice.

5.4.2.1. *Returns of use within the Rizomm*

We will successively address the lighting, heating and water supply of the toilets, specifying for each aspect the operation principles, the quality of use experiences, and finally what can be said from an analytical point of view.

5.4.2.1.1. The socio-technical scenario for office and workroom lighting

How does it work?

Within the field of lighting, sensors occupy the most important socio-technical function. Apart from circulation areas which have regulations for public buildings to require permanent lighting (in the corridors and stairways), all lighting in the offices and work rooms are controlled by sensors. They control how LED ramps are triggered as soon as entry is detected, before adjusting their intensity according to the natural light levels, by setting themselves at the standard level of 500 lux. Once the occupant has left their workspace, lighting switches off automatically after 20 min. The objective of using this device is to create lighting that is conducive to work, that is compliant, and which cuts back on energy draws as much as possible, all without human intervention.

What feedback has been gathered on conditions of use?

The socio-technical scenario is very exclusive to say the least, with fully automated and regulated management according to unavoidable criteria: regulations, attendance and departures, lesson schedules in teaching spaces. Learning about this operation was carried out in context, without prior explanation or notice on site. This was undoubtably regrettable, although for the most part the operation was quickly understood (with automatic ignitions at the building entrance, an absence of switches, a variation of intensity levels according to sunlight). How long it took for lights to switch off after the occupants departed the workspace was less noticeable. The facility manager, whose mission is to ensure quality of use and comfort, was asked to take care of the lighting and to make detailed adjustments on a case-by-case basis, and not to provide explanations as we will see later.

Overall automaticity, which has here been prioritized, has several elements worth highlighting:

– It reflects the sincere belief in the energy efficiency possible when sensors and presence are combined. This echoes the general impact of occupation on consumption levels which was mentioned earlier (section 5.2.2). Except that in this case it is no longer a question of an observation, rather a socio-technical principle that disempowers the occupant (we could even say it infantilizes them) when light sources are activated.

– Automaticity is designed to release the mental burden of lighting management, and more concretely to avoid using lighting for nothing. When taking control of the system, it is necessary to define a countdown for when the lights switch off. A 20-min interval was decided for the Rizomm building, which means that under the pretext of avoiding lighting spaces for no reason (Where? With what frequency? With what degree of certainty?), offices and workspaces are guaranteed to be lit for 20 min. without any occupants.

– The choice to adjust light intensity to the standard level means a double guarantee of conformity and comfort. However, we know that when left to their own devices, practices tend to be rather varied (ceiling light, halogen, desk lamp, exclusively natural light, occultation), as we think about specific visual comfort preferences, sensitivity to light, the construction of conditions that are conducive to personal concentration, the avoidance of reflection on computer screens, etc. Alongside this, the fact that the facility manager was asked to reduce the light intensity of workspaces because of GTB more often, attests to a diversity of preferences, even though the initial service instruction was to not respond to specific requests. Standardizing light intensity was favored, at the expense of personal expectations of occupants for suitable working conditions.

– Total automaticity also marks a desire to guard against personal sobriety, which varies according to the times and the occupants, in particular the latter, who do not pay the bills. However, in the case of the university campus, this risk has not been validated in surveys (questionnaires, observations, surveys of housekeepers or guards on their findings). It is rather only a negative representation of the energetic alertness of the occupants.

What conclusion can be made about the Rizomm building?

Analyzing the sociotechnics of lighting has highlighted two strong limits in the case of the Rizomm building. The first is the failure to take into account the diversity of habits and expectations when it comes to something that affects both well-being and productivity. The second concerns the significant (perhaps excessive) belief in the ability of the TBM-sensors association to reconcile comfort and reduced consumption. We now know that these limits were perceived by the facility manager and reported to the campus maintenance services. In this way reintroducing switches in all offices and meeting rooms has already been decided upon. Occupants will ultimately be able to adjust their lighting ambiance to their needs and preferences.

From our point of view, it is still a case of reducing the *timing* of lighting switch offs in workspaces after all occupants have departed. A sequence of tests and financial estimates on the avoided costs should make it possible to optimize the settings.

5.4.2.1.2. The socio-technical scenario of heating

How does it work?

Most of the heating device is based on the TBM coupled with the presence sensors. For offices and workspaces, the basic principle is to vary programmed temperatures, from 16°C at night and 19°C from 7 A.M. until 9 P.M. throughout the building. Presence sensors used for lighting also raise the temperature to 21°C upon detection. Then, once they start working, the occupant can request more or less heating over 4°C using a knob on the thermostat that controls their space (+/–2°C, i.e., from 19 to 23°C). Returning to the set point of 19°C takes place after 20 min. without detecting a presence, the same as for the lighting. Two devices have been introduced to limit wasted consumption: all of the doors in the workspaces are fitted with a groom which forces them to close, unless they are deliberately blocked. All of the windows are fitted with a heating cut-out device in the event of opening.

Final details about the thermostat are follows: it permanently displays the ambient temperature, which varies up to the target temperature when the adjustment wheel is activated. A small icon indicates that the heating has been switched off when the window is opened.

What feedback has been gathered on conditions of use?

Combined with chronomanagement via the TBM, decentralized settings via the thermostat and an anti-waste circuit breaker, this socio-technical system is the most complex one within the Rizomm building. Its configuration produces at least a good level of comfort in winter, produced by the default settings or as desired by the occupants. This hypothesis is all the more plausible since the programmed 21°C suggests an intention for workers in static work positions to feel comfortable. However, the available information shows that this is not really the case:

– Analysis of the manipulation of decentralized settings by TBM highlights a low rate of use depending on the floor (from 10% to 20%), which is for the most part on the rise. Does this mean that the thermal performance of the building with chronomanagement is enough to produce comfort and satisfaction for 80%–90% of salaried occupants?

– The question posed about thermal comfort attests to a clear sense of dissatisfaction: the majority of people are "not satisfied" in winter (13/22) as in summer (15/22), and "never satisfied" (10/22) are twice as numerous as the "always satisfied" (4/22).

In light of the scale of the building's renovation, the attention that is paid explicitly to the occupants (comfort and involvement) and the socio-technical system described, the results seem paradoxical and significantly mixed. A first explanatory factor is provided by the observation of the occupation conditions within the building. That is to say, the high performance attached to the renovation has led to heat production scenarios that are associated with occupancy formats which are far from reality. For example, a large office for two or three people can be equipped with the same heat emitter as a small office for one person with the rationale that a higher occupancy is favorable to indirect heat production (through computers and body heat). However, the individualization of professional constraints, not to mention the possible variation in hours of sunlight, means that co-habitation in the same space is more the exception than the rule. We have also seen unwise choices when it comes to technical devices, as is the case with blackout blinds that are identical everywhere and insufficient to protect south-facing workspaces and offices. Even before any intervention by the occupants, the building equipment and the modeling of presences cause a level of discomfort in summer and winter which will have to be suffered or corrected by socio-technical "constructions". A lack of satisfaction and the lack of use of socio-technical elements (a decentralized thermostat, a facility manager, a groom, blackout blinds) are proof that something did not work.

What conclusion can be made about the heating of the Rizomm building?

Comfort expectations have been raised due to the fact that the campus has been shown to be innovative and the renovation of the Rizomm has helped it to become a demonstrator building. Yet, in this particular context, the occupants' low satisfaction is in no way a sign of failure; it is rather symptomatic of the great difficulty it takes to reconcile energy performance and comfort.

To illustrate this, it suffices to assume that consumption and/or cost limits are absent: the rise in temperatures means that the coldest people end up feeling hot, while those who are overheating lower their heat emitters if they can or, in the more likely scenario, they mitigate overheating by opening windows. Here, everyone can feel satisfied, but this comes at the cost of significant bills and GHG emissions. The case of the Rizomm building is different. Here, overall performance (energy and comfort) is neither simple nor instantaneous. It is still yet to be found.

In view of what has been put forward regarding the socio-technical functioning of a building, the issue of discomfort related to the Rizomm building can be compared to three parameters which are also avenues for correction:

– It is imperative to correct the imaginary of efficiency which is attached to buildings that are proclaimed to be efficient, in order to better underline the essential contribution made by occupants. For example, it would be wise to illustrate the reality of the socio-technical mechanisms that are inherent in "constructing comfort" so as to show that the occupant develops and cultivates a thermal environment by playing on the devices within their reach and with the tricks that they have come to know, most often without even realizing it. Even though it is very autonomous, an efficient building does not undermine how relevant this contribution actually is, since it operates via default adjustment parameters or according to needs.

– If we were to speak more of the occupant's contribution, we would expect greater attention made to the socio-technical resources made available to them. A priority would be to correct the lack of knowledge around the potential for thermal construction provided by an assembly of groom-thermostats and decentralized-blackout blinds. Low use of the thermostat temperature dial highlights a lack of education, despite the presence of the energy manager.

– It is imperative to know as far in advance as possible how exactly the spaces are occupied (occupation time, actual densities of the presences, modification of the purpose of the spaces, etc.) in order to adjust the GTB as effectively as possible. Like with any building, the Rizomm hosts activities that vary in time and by nature. It is imperative to provide a mechanism for monitoring these variations using a logic of continuous adjustment.

The Rizomm is a good illustration of the need to not equate the search for energy performance with that of thermal comfort, a criterion that determines the working conditions of occupants.

5.4.2.1.3. The socio-technical scenario for supplying sanitary facilities with water and lighting

How does it work?

The toilets are supplied with sensors: one in the sink area (lighting + for toilet flushes), and the others in each toilet stall (lighting). The control sensors for lighting offer a common device; several minutes are programmed before the lights go out in the absence of a new presence detection. The toilet water supply also works using sensors to prevent losses by leakage or blockages. The TBM could allow for a rapid identification of losses, provided there is an alert system, which is not yet in place.

What feedback has been gathered about conditions of use?

The triggering of the lights at an entrance is a system that has been well received by the occupants, provided that they do not switch off too quickly in the toilets and that the position of the sensor allows for a new detection. This last point will vary depending on the size of the toilet. How sensors are positioned does not pose any particular problem in the vast majority of cases. On the other hand, the system that is put in place for water is less satisfactory. The cause of this is that the location of the sensor which is placed at the entrance of the sanitary space only offers a single detection time. However, in various circumstances, the activation of a single flush may not be sufficient. The user is then faced with three possible strategies: wait for a new person to be detected to trigger the device, go out then come back to the toilets to activate the presence detector, or simply leave the toilets as they are.

What conclusion can be made for the Rizomm toilets?

When it comes to the use of sensors, one can quickly identify limits in their effectiveness: in terms of detection (simple?), timing and intensity of the settings (wise?) and adaptation to different scenarios (flexibility?). Within the sanitary spaces in question, it was possible to point out imperfections in each device:

– In terms of lighting, occupant satisfaction depends first of all on how long lighting was switched on for, which only feedback from use can correctly assess. It then depends on ease of detection, which is linked to the location of the detector in the toilet. The solution would be either to review the positioning on a case-by-case basis, or to increase the sensitivity of the sensor, which is always placed on the ceiling to allow for wired connection to the network.

– The socio-technical water system marks the distinction between a theory, inspired by an intention to avoid waste, and real needs. Critically analyzing the uses allows one to identify a problem when it is not so much about making an adjustment as it is about doubling up detection so that it covers the entrance to the toilets and presence in the toilets. It would also be possible to move away from a purely technical solution, for example, by setting up a system which allows one to alert and/or report an inconvenience to the technical services.

5.4.2.2. *Returning to use within the HEI building*

The preference for full automation in this building reduces the margin of intervention on the part of the occupants within their environment, but does not eliminate it altogether. As has already been explained in particular with regard to thermal comfort, the occupants are never passive when faced with a situation that does not suit them. Studies about the sociology of energy report numerous initiatives and instance of ingenuity which prove their activism.

Initiating a return to use the HEI building is a new opportunity to prove the non-passivity of the occupants, despite the related socio-technical choices and the more or less explicit request for non-direct intervention in the facilities. For this, we can look at a survey which was carried out at the beginning of 2020 based on questionnaires, field observations and spontaneous exchanges with the occupants. The objective was to measure the satisfaction level when it comes to the comfort experienced in the workspaces (mostly open spaces) as well as relaxation. This survey was the first in a process planned to last in order to verify the expected progress over time, until the whole building is converted into a smart building.

The following will be addressed successively:

– quality of thermal comfort, excluding sectors with faulty insulation;

– ambient air quality, ensured by dual-flow air handling units (CO_2 limit value set at 800 ppm);

– quality of the lighting environment.

5.4.2.2.1. The perceived quality of thermal comfort

How does it work?

The heating works using a hot water circuit supplied by an urban heating network. The programmed set point temperature is 20°C, a strict minimum due to the static postures inherent in office or student work environment. This set point is often increased in winter throughout the building to take account of poor insulation as a result of poor workmanship in certain well-marked sectors. This is a big problem for occupants and a challenge for managers in the absence of basic functioning. To moderate consumption, the set temperature is combined with the presence schedule in the rooms that have one person as well as the number of people planned to occupy the space.

What feedback has been gathered about the conditions of use?

The general tone of the thermal assessments is rather negative. At best, half of salaried occupants say they are satisfied with the temperature in the office, and the others complain of temperatures that are too low. Dissatisfaction is more pronounced among students who mainly sit in classrooms and lecture halls.

The thermal discomfort expressed has two characteristics:

– Zoning, with variations between floors which are sometimes close to 2°C, affecting circulation areas, offices and collective workspaces. The feature is reinforced by gaps within the spaces frequented, by the proximity of windows, ventilation openings or doors. This parameter is reflected in terms of how differences are perceived by three occupant profiles:

- according to their location in the building: teachers/students, administration/ researchers, between groups of researchers according to their specialty;

- according to the nature of their work, with varying levels of stasis with vulnerability to ambient cold;

- according to their gender, with women being more critical, because they are also overrepresented in administrative functions and due to their physiology, which is more sensitive to low temperatures.

– Seasonal irregularity, with variations between morning and afternoon, and often over the course of the same week. The origin is above all technical in nature, with a difficult balance between the supply of heat and ventilation; this last problem arises above all in classrooms and lecture halls, which are more exposed to the deterioration of air quality due to a concentration of students. A student expresses: "It is relatively unpleasant to have to change clothes depending on the room where we have class".

Thermal "construction" strategies to protect oneself in spaces that are too cold fall within a fairly standard set of actions. Layering up on clothing, before putting the clothing intended for going outside back on, is the easiest and most widespread solution. But because the discomfort experienced is frequent, the norm is to be equipped with an electric heater as this will immediately provide a lasting supply of heat. Other solutions are identified, which are more punctual (consuming hot drinks) or too radical to be taken on by all (changing area or room).

The best illustration of the sequence between a perceived lack of comfort, the construction of a thermal solution and the costs incurred is provided by the testimony given here. It attests to the non-passivity of the occupants who not only seek comfort in their personal workspace, but also project themselves into how they function across the day, with the necessary circulations and internalized thermal zones which dampen any effects. There is consideration here for the energy and financial effects of these behaviors; the legitimate priority is indeed comfort at work and the preservation of one's health: "It's cold when you arrive in the morning [...] in the corridors [...], hence a thermal shock when leaving the office. To avoid this, open the doors of the office while heating it. When we do this we experience the inconvenience of possibly hearing noise (of people passing by)".

What conclusion can be made about the HEI building?

Above all, it should be kept in mind that thermal assessments were collected in a building that is still technically being finalized and is in the process of discovering the results obtained for given settings. However, since the general design of the socio-technical heating system and its operating principles have been well established, the trends identified can be commented on.

Searching for the thermal performance of the building (temperature and consumption) is mainly based on a triptych: a set temperature, timing of the temperatures according to an occupation schedule in the classrooms and workspaces, and presence sensors to check actual room occupancy. Confronted with the actual conditions of implementation, these options produce moderate satisfaction among occupants, due to a lack of consideration of the building parameters (thermal weaknesses), occupation parameters (work with varying levels of mobility, gender) and the difficulty of reconciling comfort performance (heating and ventilation). Faced with this observation, it is necessary to question the capacity of the intelligence expected in the building to overcome these sources of weaknesses or to question whether there is a need to modify the performance paradigm put in place.

The answer to this question cannot be categorical. Despite everything, it is obvious that the variety and interconnection of dissatisfaction factors invite us to think in negative terms. Indeed, the modes of occupying spaces and the factors that produce a level of nervousness are so varied and ever-changing that they would justify micro-adjustments when it comes to occupants (duration of presence, nature of the work, sex, personal sensitivity, etc.) and not just general settings which can even be associated with security, such as sensors to verify effective presence in the spaces that have been announced as "occupied" for chronomanagement.

The 2020 Comfort Assessment Report suggested scalable and adaptive comfort (or temperature) settings. The functionalities of the TBM should make it possible to improve adjustments to needs, on the triple condition that they are known in detail, that knowledge has been updated and that technical services agree to spend time modifying the TBM in result. Another solution would be to introduce a degree of decentralization in the settings according to a reference that would still have to be defined: the room, the type of room, the floor, the sector, etc. This ultimate solution would also be the way to respond to an occupant's expectation to control the environment in tertiary buildings [CER 13].

5.4.2.2.2. The quality of the lighting atmosphere

How does it work?

The lighting system operates using presence sensors. The lights turn on according to an intensity that is based on the requirements of French regulations. If it does not vary according to exterior light intensity, the occupants are able to switch it off using switches. The lamps are LEDs in our reference building.

What feedback has been gathered about conditions of use?

The satisfaction about lighting is good in the HEI building, with a rate ranging from 64% to 83% among employees and from 85% to 99% among students. These two populations do not frequent the same areas. Three circumstances cause satisfaction to travel downward:

– low exposure to natural light, through poor orientation and/or small windows. The comments relate more to regret and psychological discomfort;

– excessively strong artificial light, especially when low ceilings make it difficult to diffuse light. The occupants mention problems surrounding glare, concentration, headaches;

– the operation of sensors is not considered optimal, due to a positioning problem and/or lack of sensitivity to presence.

Faced with a situation of visual discomfort, the range of possible solutions appears less extensive than when it comes to thermal discomfort. In any case, this is what the comparison of passivity rates suggests (50% of salaried occupants against 2% faced with thermal discomfort). When actions are undertaken, they are above all radical in nature, such as a change of location in the occupied space (going to another spot) or in the building (going to another place). Employees who frequent a personalized workspace have access to specific levers, such as the manipulation of curtains or the use of an extra light in order to create a more intimate atmosphere in the office.

What conclusion can be made about the lighting atmosphere quality at HEI?

Judgments about working conditions are always captured according to three modes: too much, enough or not enough. In terms of lighting, the HEI building has a good satisfaction rate but suffers from a lack of natural light in disadvantaged areas, a poor positioning of sensors and, more frequently, a light intensity that is sometimes too strong. The results are therefore rather positive on this criterion.

There are obvious correction possibilities because of future TBM, on the initial functional condition of wanting to take the occupant's point of view into account, when it comes to adjusting the intensities, almost on a space-by-space basis. It is understandable that proper technical operation and compliance with regulations take precedence over technical services. Yet, it is still necessary for occupants' perspectives to be collected (which the survey allows for) and for them to tell their own account.

5.4.2.2.3. Ambient air quality

How does it work?

The air in the building is renewed by treatment plants, with a ceiling value of CO_2 of 800 parts per million (ppm). Technical services accumulate data in order to be able to change the instructions according to the number of students planned per course. For now, CO_2 is the only parameter that is taken into account by the sensors.

What feedback has been given on the conditions of use?

Air quality is an aspect of comfort for which distinguishing the points of view of employees and students makes the most sense, for the question of spaces that are frequented and human density that is experienced. These two populations are exposed to different degradation factors in their own ways, namely:

– human concentration, with a rapid risk of CO_2 concentration and body odor;

– equipping spaces, according to decoration materials and furniture, the presence of supplies or archives;

– the possibility of manual ventilation, if the window is accessible and able to be opened.

Overall, air quality assessments vary according to the exposure of each space to one or more of these factors. They are rather positive (about 80% of “good” or “fairly good” perceptions) with lower scores in five location profiles:

– offices exposed to dry, heavy air, with strong smells and dust;

– meeting rooms without windows;

– classrooms or labs caused by a concentration of human bodies;

– computer rooms, which require cooling;

– employee break rooms, with windows that are not easily accessible.

The effects of a deterioration of ambient air are varied: headaches, loss of concentration, dry nostrils, feeling of suffocation, etc. However, ventilation does not work, sometimes to a large degree, with almost 20% of students complaining of a flow of cold air and/or noise in computer rooms.

There are not so many ways to solve an ambient air problem. The equipment is impersonal and harbors anxiety-provoking springs (can and should we intervene on a ventilation system?). Solutions turn out to be quite disjointed according to the two occupying populations: employees adopt the opening of the windows because they are in the offices (55%), passivity (25%) and marginally withdrawal (10%). Students are much more inactive (45%), they change places (36%) or they request actions of a third party (20%).

What conclusion can be made about air quality at HEI?

Air quality is a theme that is more present in concerns, even if it is still secondary in comparison with thermal comfort. Occupants act at home and at work with the same concern for ambient air renewal and for getting rid of unpleasant odors. Opening a window as a first port of call to be adopted by employees is logical and effective when possible. As for students, we hypothesize that their greater passivity is only due to their status (they are supervised and not direct decision makers) as well as to a certain social laziness which results from the group effect.

The choice to equip windows with a heating cut-out system is a sign that manual openings are today perceived as a loss of thermal efficiency and that they are undoubtedly bound to become rarer because of consumption control. It is difficult to take a position on whether this prospect is wise or not. We only know that equipment has a flaw as it makes building security problems visible (through bad weather, intrusions), which requires time-consuming interventions for security personnel. On the other hand, it is necessary to once again draw attention to the point of view of occupants. Doing this will make it possible to identify the deterioration factors that must be dealt with in one way or another: a change of layout, grouping recent archives to empty offices, diverting ventilation outlets to avoid noise and cost to the occupants, continuously adjusting airflows according to CO_2 and occupancy density, etc.

Air quality also has little-known operating and adjustment principles. Insofar as this quality (or its degradation) is less perceptible than thermal comfort, it is also essential to promote pedagogy so that everyone is informed about the design of the device and the adjustments that are accessible to the occupants.

5.5. Lessons to be learnt from the two pieces of feedback

After at least one year of occupation and despite the technical hazards that may have disrupted the smooth running of the facilities, it is possible to assess the socio-technical functioning of the two renovated smart buildings. The interest here is to provide general lessons based on the effects of socio-technical biases by item of use. The points listed are to be considered without any hierarchy and from a systemic perspective: the technical aspects and uses are very interdependent.

5.5.1. *Do not confuse intelligence, performance and quality of use*

5.5.1.1. *Adapt the degree of automation to the building*

The HEI and Rizomm buildings are a reflection of the growing smart building market: there will be more and more old buildings converted due to available stock in the existing stock. From this perspective, we must prioritize questioning the performance paradigm which is the most suitable, the degree of automation of the settings being a pivot point. The analyses and findings put forward point to advocating for restraint in this regard.

Before considering personal needs and preferences, the occupant's "construction" activity is encouraged in the context of a building which a renovation will always struggle to resolve (differences in orientation, thermal fragility due to choices made by the architect, roof layout, etc.). The renovation–intelligence combination makes it possible to reduce the comfort problems that are experienced by the occupants, but not to eliminate them. In any case, there is a strong trend that emerges from the research work and a deep dive into the two rounds of feedbacks from use described above. Consequently, preserving a margin of maneuver for users, however small, appears to be the most realistic socio-technical option for achieving the highest degree of comfort. Performance in the modulation is to be sought by taking into account the context of use, rather than in the generalization of a socio-technical bias based on the technical whole.

5.5.1.2. *Remaining vigilant about the role and operation of sensors*

Sensors have become an emblematic piece of equipment in high-performance buildings. Embodying what may correspond to a "non-human actor" [LAT 91] in an almost ideal typical way, they track events and circumstances that require technical devices to be modulated, upward or downward. Their presence is also a clear sign that "human actors" have been sidelined, in perception, by decision or due to user preferences. The two buildings are not far from displaying extensive sensor instrumentation, insofar as all or part of the consumption stations that are in contact with the occupants are equipped with them. This observation of the situation, which reveals a major trend within so-called "intelligent" buildings, justifies the reality of always having to ask a series of questions prior to implementing the equipment: What is the reason behind implementing them? How much will they be used? What do people think about the presence of sensors? These are devices which have their limits in terms of functionality and reliability, and they have their own requirements to be properly operated. So, there are many essential questions to ask before making any adoption decision.

5.5.1.3. *A good measure of intelligence through usage performance*

For a smart building, performance is expressed in terms of energy and quality of use. However, we know that beyond energy performance which is linked to an improvement in fundamental thermal energy (insulation, ventilation, efficiency of the installations), energy performance is always very difficult to assess due to possible technical hazards, as was the case for our buildings, to which possible variations in activity may be added. Within this logic, quality of use is essential as a major criterion for performance, and this is true however the activity or the forms of occupation evolve in the building. The quality of use does not imply special circumstances: users must always evolve in the best possible socio-technical conditions when it comes to regulations, to maintain well-being, safety, and most certainly for energy performance, since the acts of "construction" equalize or add to energy consumption more than they reduce it.

5.5.2. *Having an accurate understanding of occupations and uses*

Since the quality of use is to be considered as the major criterion of performance, including in a smart building, its measurement must cover the different facets of occupancy. The challenge is putting building services and user needs in perfect harmony.

5.5.2.1. *A detailed breakdown of the activity system*

One first needs to follow how the contents of the building evolve in terms of activity. The uses are determined above all by a purpose and an activity system, which entails physical postures, temperature needs, relations between colleagues of various levels of density, equipment whose uses may or may not be programmed, etc. This approach is micro-organizational in nature (how is the work of each peer organized?) but not yet micro-individual (how does each individual organize their work?). More objectively, adjustments are facilitated before the launch of a building, which then makes it possible to understand or even objectify the socio-technical adjustments that need to be made in order to stick to the functional requirements.

5.5.2.2. *Do not equate occupation and uses*

Programmed and/or effective presence is the main mode of regulating energy intensities in smart buildings equipped with sensors, as is the case in our two buildings. But this socio-technical principle is notoriously insufficient when it comes to effectively adhering to usage.

On the one hand, occupation varies in density thus creating variations in ambient temperature, with a greater risk of thermal dissatisfaction in the event of lower density. This can be verified in lecture halls as well as in the offices of two or four people. On the other hand, physiological needs vary between people. In these circumstances, the attention paid to the quality of use requires that the occupancy criterion be enhanced with a modulation device that properly matches needs, other than a trigger associated with a set temperature. Buildings need to incorporate control modes that can work for various scenarios or use cases, that is, they need to be adaptive and scalable.

5.5.3. *Informing users about the importance of their role*

Research carried out on the energy consumption of buildings has shown that on average 20% depends on the occupants. This share is less than that related to technical aspects, but it nevertheless represents a major challenge in terms of the environment and bills to be paid. Added to this is the issue of the quality of use, which only the occupants are able to define and adjust within the work context they experience, whatever the nature of the building. Mobilizing occupants therefore becomes a necessity. The key word to sum up their contribution is "optimization". This can perhaps be carried out technically, but it would be without reaching the qualitative intensity that is provided by the experiences of occupants in a work situation.

5.5.3.1. *At the service of reducing energy consumption*

The intelligence of a building does not change the situation: the occupants are key players for energy performance because they control two major uncertainties. Apart from ambient energy services (heating, lighting, ventilation), they are first and foremost the direct users of energy at work, since they control its periods, intensities and changes according to activity requirements. They also decide on the methods of use, which have possible flaws if the instructions for use are poorly followed or if people are ignorant about how the devices function. As building occupants, they also know how they intimately work on at least two levels: the quality of adjustment between settings and expectations (set points, intensity, chronomanagement, etc.), and in terms of malfunctions (faults, adjustments).

5.5.3.2. *At the service of quality of use*

For the same reasons as for energy consumption, occupants are in the best position to assess the quality of the service provided by the settings in a building and to suggest areas for improvement. We know that they act spontaneously to increase their level of comfort, either respecting the technical prescriptions or not. This will sometimes have negative repercussions on energy consumption. But the issue of overall performance (energy and quality of use) requires attitudes and practices for improvement to be made more visible in a logic of use, in order to appreciate the positive and negative effects that are induced.

5.5.4. *Developing organizational regulations*

The description of the socio-technical modes of operation has insisted on the importance of the contribution of the occupants, including and perhaps even more so in the buildings that claim to be efficient. This is a paradox of the current evolution that was not anticipated nor revealed by sociological work: the more demanding the expected results, the greater the expected contribution of the occupants, whether it be total passivity that is demanded (as in the HEI model) or a clear promotion of human contributions (as in the Rizomm building). Either scenario requires understanding, assimilation and a minimum level of cooperation on the part of the occupants, which is not something that comes spontaneously.

5.5.4.1. *Socio-technical engineering, a profession of the future*

The inherent complexity of high-performance buildings poses problems related to understanding for the occupants. The devices within their reach do not always operate intuitively and easily (quality of appropriability). The presence of sensors is intriguing, and it is difficult to know what they control and whether they can be manipulated, adjusted or moved without risk. All this ambient complexity needs to be deconstructed so that uses are at least compliant, or better still, they produce a legitimate level of comfort which is sought by users, without compromise. Added to this is the increasingly frequent presence of a TBM. In a logic of technical management, this equipment constitutes a valuable resource for making the essential adjustments to respect what is prescribed during the design phase, before refining this in order to meet the expectations of the users as and when they pop up. The existence of all these imperatives justifies the development of a socio-technical engineering function, positioned at the functional interface between the socio-technical environment and the occupants. Provided by a facility manager, the position is suitable for a person with two skills which are of equal importance:

– one technical, to intervene on the GTB and follow the evolution of the devices;

– the other relational, to explain the philosophy of the building, the limits and potential of its equipment to users and new arrivals, and to collect suggestions for adjusting the TBM.

The facility manager is the socio-technical regulation lever who is essential to any smart building. They also become the actor through which the technical services, which up until now have favored the absence of breakdowns as a performance criterion, can make their business culture evolve.

5.5.4.2. *Involving occupants in socio-technical devices*

The role played by the occupants of buildings has been valued on many occasions, with regard to energy consumption and in order to improve the overall quality of use. Their involvement was above all presented as taking the form of relations that were not very formalized, apart from the informational times provided by the technical services to explain the operation of the technical devices, as well as the procedure for reporting and then dealing with hazards.

It is useful to specify that involvement can take more organizational and official forms. One may think in particular of the occupants who become the "energy correspondents" of their building. Their role is to become a link between the technical services and the community of users. The main interest is to bring out consensus when it comes to the malfunctions that need to be resolved as a priority, to form part of ideas for improvement.

5.6. Conclusion

Usage performance is certainly more difficult to obtain than energy performance, because the human factor cannot fit into any scientific law. To achieve this, the sociology of energy favors the promotion of user involvement as early as possible, in the decision-making process of the building philosophy, then in that of the design of socio-technical devices. This has been done without much success so far, because what is demanded takes time, costs money and does not claim to provide the guarantee for an effective and positive response. Exploring the dynamics of use and the operating methods of two buildings offers some encouraging signs for the future and a strong consideration.

First, we can see that a contradiction that is becoming accentuated: between calls for greater responsibility and commitment of energy users, a growing level of sensitivity and commitment to the environment within the population, and a perpetuation of social distancing in working environments. In the somewhat long term, we must hope for a paradigm shift which increases the involvement of users. Second, the slow but real development of socio-technical engineering, embodied by the profession of facility manager, demonstrates a concern for a new dialogue with users. The future will tell us whether this organizational change will be conducive to a more frank consideration of user points of view.

Finally, a strong observation relates to smart users. The advent of smart buildings and changes in the socio-technical context that this implies have revived the issue of taking the user into account. The description of the dynamics of use and the two feedback loops show that the occupants are always as quick to interact with a socio-technical environment with varying levels of automation, to increase their feeling of comfort and reduce their mental load. Beyond this observation, it should be noted that the exploratory work that has been carried out has made it possible to advance the understanding of what a smart user is or should be. Unlike the occupant acting to increase their comfort and according to their free will, the smart user would be an "enlightened" actor, that is to say, informed about the modes of use to be respected and the challenges of energy control, and "cooperative". Always free in their actions, they would know when to inform technical maintenance of their practices. The powerful and appealing technophile imaginations that are associated with smart buildings tend to accentuate how an occupant to be neutralized is represented. The sociological analysis that is applied to smart buildings leads us to think quite the opposite.

5.7. Acknowledgments

The renovation and transformation of the Rizomm building into an experimental building benefited from funding from the Hauts-de-France region as well as European ERDF funding (European Regional Development Fund); the transformation of the HEI building in Junia into an experimental smart building benefited from funding from the Hauts-de-France region as part of the Live TREE program.

6

Ethics of Energy and Societal Transition

"Natural, technical and social sciences can provide essential information and evidence needed for decisions on what constitutes 'dangerous anthropogenic interference with the climate system'. At the same time, such decisions are value judgements."

[IPC 01]

6.1. Introduction: ethical challenges associated with the energy and societal transition

A set of facts seems to show that climate change is now well anchored in our global reality and that its effects are and will continue to be particularly devastating for the planet, biodiversity and the populations that inhabit it. Ways of life and consumption of human beings are key stakeholders in these changes. As there seems to be a consensus to say that this poses a problem for humanity now or in the future, it is necessary to question why. Is it because a sustainable environment is essential for the survival of humanity? Or is it because nature and the different elements that constitute it have a special value that must be respected and protected? These types of questions arise not only for individuals (What should I do with my waste? Should I reduce my car travel?), but also for society as a whole and the institutions that organize themselves around it (What waste treatment policy should be implemented? How does one reduce energy consumption?).

Consequently, the energy transition as a "structural and profound modification of the modes of production and consumption of energy" [LAV 18] can only be accompanied by societal transition that involves the modes of operation, or the "ways of life", inherent in our societies. Moreover, the energy transition cannot be thought about independently of the relationship that mankind maintains with nature.

Any ethical reflection on these questions can only be evaluated in relation to actions: what are the ethically justified actions that make it possible to work in favor of this transition?

We can already see the links between the transition and reflections on the attitude we should maintain when it comes to the environment. The first ethical question that may then emerge concerns an articulation of the individual and the institutions in charge of organizing the life of the societies in question. This can be conceived in terms of responsibility and obligation.

Ethics, in a simple and fundamental way, seeks to define the criteria to distinguish between good and evil, and more precisely to distinguish good from bad actions, a good action being obligatory and a bad action being forbidden. Therefore, one can very quickly link ethics to the question of the justification of actions. Thus, the question can become whether is it possible to justify moral obligation[1] in order to act in favor of an energy and societal transition[2]?

However, it is clear that at least *four ethical challenges* are raised [BRO 12]. The first lies in *how to act in a situation of risk and/or uncertainty*. The second consists of defining the way in which it is possible to *compare damages and benefits that are spread out over time*. Indeed, on the one hand, climate change will persist over time while on the other, the measures that can be taken today will take several years to have any real effects. How does one therefore ethically assess the need to make major investments today to fight climate change, when the effects will take place in the future? In other words, how can we ethically establish an immediate sense of obligation to gain future benefits, considering the fact that the people on whom this obligation falls will not themselves benefit from it? The third challenge raises the question of the value that is to be attributed to human lives. Thousands, if not more likely millions of people, will be victims of this situation. They will suffer from the effects of climate change whether it is due to famines, heat waves or even reduced water resources. How do we take into account and define the value of each human life involved? Finally, the fourth challenge concerns population ethics. As the effects that we have just mentioned will have an impact on the world population, how then should we think how it will evolve? How do we integrate the possibility of global catastrophe into our ethical reasoning?

These four points alone represent immense ethical challenges; they also reveal three important impacts on the way any ethical proposal should be designed [MUL 11] which could justify action in favor of the energy transition. First, *the*

1 Here, we will not make a distinction between ethics and morality.

2 We limit ourselves here to a procedural or ideal reflection that aims to understand the nature of the arguments that can be advanced without giving it specific substantial content.

conflict of interests between people living now and those who will live in the future becomes real. Next, the prospect of a less prosperous future and/or with living conditions different from those of today *forces us to reassess what we understand by a "good life" or a "flourishing life"*. Finally, if we can legitimately assume that the *basic needs of the majority of the world population cannot be met in the future, then we must face tragic situations where the choice between life and death* will no longer be simply hypothetical. In other words, thinking about the ethics of energy and societal transition involves *rethinking and questioning three presuppositions* that have structured modern ethical thinking: *the interests of present generations coincide with those of future generations, people living in the future will be better off to the extent that mankind and our favorable living conditions will continue indefinitely, and the people of the future should therefore not be considered in any type of ethical reflection.*

Consequently, to develop an ethical reflection on the energy and societal transition, one must first establish the moral obligation to work in favor of this transition and therefore of protecting the environment. Once this obligation is established and given that it can only take on meaning through action, the question arises as to the consistency between recognizing such an obligation and carrying out the action itself. However, we can see that there is often a gap between the action we consider to be ethical (under ideal conditions) and the action we perform (under real conditions). The question is no longer about obtaining a justified evaluation, rather about *the way in which we can influence individual behavior*, particularly through public policies that target the greatest number of people. However, we have witnessed a major turning point in recent years: nudges have emerged. They are a form of soft paternalism which aim to influence the behavior of individuals by changing the context in which they make decisions and they are based on behavioral sciences insights. The latter have the particularity of being able to offer a vision of the individual that relies on cognitive biases, a kind of thought pattern which causes the cognitive processing of information, leading a person to make predictably irrational decisions. Finally, beyond their effectiveness (real or supposed), it is necessary to ask questions about their ethicality and their legitimacy. We will then be able to evaluate nudges as public policy tools that promote the energy and societal transition. This chapter therefore endeavors, on the one hand, to ethically ground the very idea of energy and societal transition and, on the other hand, to evaluate a very specific public policy tool: the nudge.

6.2. Some arguments in favor of the energy and societal transition

It seems to us that a necessary precursor to our reflection consists of identifying the nature of the arguments that can be mobilized in favor of the energy and societal transition. For this, we must face a certain number of questions: how should we think about the ethical relationship that human beings must maintain with their environment? How can we explain the strong reluctance of some people to take the effects of human action on the environment into consideration? Conversely, does taking these effects into account necessarily lead to a specific ethical consideration regarding the environment? Warnock summarizes this type of question in purely ethical terms:

> Let us consider the question to whom principles of morality apply from, so to speak, the other end-from the standpoint not of the agent, but of the "patient". What, we may ask here, is the condition of moral *relevance*? What is the condition of having a claim to be *considered*, by rational agents to whom moral principles apply [WAR 20].

6.2.1. *Assign a value to the environment*[3]

In this way, it would be possible to defend the position that the relationship of mankind to nature is based on a principle of "free use". In other words, natural resources are the primary means of satisfying human wants and needs. Talking about environmental protection makes no sense and can in no way establish any moral obligation. Such a vision is anthropocentric in the sense that the relationship to nature is thought of in relation to the needs of human beings. Aristotle thus affirms that nature is made specifically for the good of men [ARI 04]. In such a conception, nature only has an instrumental value: the only value that can be assigned to it reduces it to a means used to achieve other ends. For example, some fruits have instrumental value for animals because eating them allows them to survive. If we consider a person to have instrumental value, then such a statement is always accompanied by the attribution of intrinsic value to that same person. Indeed, as a person, they have an intrinsic value independently of their instrumental value (and even if they have no instrumental value): they are themselves their own end. The distinction between instrumental value and intrinsic value is fundamental. Indeed, let us take the example of a wild plant that is used to create a drug. Its purpose is to heal a human being, so that it has an instrumental value. We could however consider that this same plant has a certain value independent of its ability to heal people, in this way we could speak of intrinsic value. Why is the question of whether a natural element has intrinsic value important? Quite simply because if this is the case, it implies that moral agents have a direct and *prima facie* moral obligation [ROS 02] to protect it or at least to refrain from damaging it [JAM 02, ONE 92].

Mankind's relationship with nature can indeed be questioned in terms of the value attributed to it. Earlier, we highlighted the anthropocentric conception. More

3 We here rely on [BRE 21].

precisely and with regard to what has just been stated, *strong* anthropocentrism offers intrinsic value to human beings only. A *weak* anthropocentrism places more value on human beings than on non-human beings or things. This leads to almost always justifying the protection or promotion of human interests, to the detriment or cost of non-human beings or things. In the same way, destructing the environment or natural resources is not condemnable in itself, but it is only because it can reduce the well-being of humanity in the short term or long term [BOO 90, NOR 95, PAS 74].

With this genre of reasoning, there is no moral obligation as such to act to preserve the environment, and by extension, there is no moral obligation to act in favor of the energy transition. Therefore, if we wish to highlight any sense of moral obligation toward nature, the environment or biodiversity and therefore ethically justify the energy transition, it seems necessary not only to *question the moral superiority granted to humanity, but also to grant an intrinsic value to nature, to the environment and to all "non-human" elements* that compose them. Such a project may seem particularly complex [DES 94, LIG 96, NOR 91]. This is why prudential anthropocentrism could be sufficient, as it would focus on the type of justification that could accompany public programs or policies that are aimed at preserving nature. The argument would then be the following: the moral duties we have toward the environment are based on those we have toward its inhabitants. What we lose in the intrinsic value given to the environment, we gain in the ability to "simply" and effectively justify policies that are aimed at protecting it, or at least at not degrading it.

This is the theoretical framework in which ethically evaluating the energy and societal transition becomes possible. The second lesson that we can learn is that the value we give to an entity confers upon it a specific moral status. Thus, granting an *intrinsic value to the environment* makes it possible to sketch out a moral obligation that justifies its preservation, independently of any other consideration. Granting it an *instrumental value* makes it possible to uncover a similar moral obligation, which will then only be considered from the point of view of preserving the well-being of human beings.

6.2.2. *Responsibility toward future generations*

In the same way that unearthing a moral obligation toward an entity that does not meet the conventional definition of a moral agent[4] and from which one cannot expect any reciprocity, thinking about responsibility toward future generations, in other words, moral agents who do not yet and perhaps never will exist, is a

4 A moral agent is a person (or entity) who is able to form moral judgments on the basis of even a rudimentary distinction between good and evil and who is therefore considered responsible for his or her actions.

challenge for any ethicist [GOS 04, MEY 21]. Relations between present and future generations are inevitably and constantly asymmetrical. However, how can we think about an energy and societal transition only by taking the interests of people living *today* into consideration? Would it even be possible to justify ignoring the long-term consequences that the actions of current generations may have on the environment, engendering a specific environmental state which will last beyond present generations and which therefore will affect future generations? William Stanley Jevons already asked this question in the 19th century in the context of coal [JEV 65].

Is it not therefore our duty to leave an environment in which future generations can live decently? Indeed, present actions can affect future generations in many ways: they can have an influence on their desires, living conditions and interests, whereas the reverse situation is not possible. More fundamentally, the very existence of future generations, their number or their identity may be at stake [MCK 17, MCM 81, MUL 15, SCH 13].

It is therefore a question of reflecting on the principles of intergenerational justice [CAN 21]. These principles can be based on a sufficientist approach, which simply considers that all people must have living conditions which are above a certain threshold – the challenge obviously being to define such a threshold [MEY 09, PAG 06, PAG 07]. Defining a "sufficient" threshold could nevertheless result in one generation leaving the planet in a state that would mean that future generations have to live in worse conditions, but conditions that still sit above this proclaimed threshold. To avoid this pitfall, Llavador, Roemer and Silvestre propose the idea of "sustainable growth", which can be defined as follows:

> Growth sustainability (say, at 25% per generation) means to find that path of economic activity that maximizes the welfare of the present generation, subject to guaranteeing that welfare grows at least at 25% per generation, forever [LLA 15].

With regard to the ecological impact of economic growth, such a conception nevertheless seems to come up against several difficulties, in particular that of the resources that are necessary for such growth. It also presupposes that future generations will inevitably be richer than present generations. An egalitarian approach also makes it possible for us to think about the demands for justice between generations. Thus, present generations may have the duty to make decisions concerning the environment and the climate that will not accentuate inequalities in future generations [CAN 18, FLE 19]. This is all the more important as we already know that, for example, climate change reinforces existing inequalities [HOE 19].

Nevertheless, a question arises here, particularly with regard to fossil fuels (but not only in this sense), which Axel Gosseries has perfectly identified:

> To what extent does the obligation to pass a basket of goods of an equivalent value to the next generation also imply that the composition of this basket must remain of the same nature? [GOS 04]

What he invites us to think about is the possibility of substitutability which, though it cannot always be complete, can admit certain degrees of difference. The limits of such substitutability lead him to think that "we must take care to preserve a certain proportion of physical capital in the basket which is transferred to the next generation" [GOS 04].

Therefore, the energy transition should probably and at the very least aim to find a balance between the needs of present generations and the availability of resources so that future generations can also benefit from them[5]. In other words, responsibility toward future generations makes it possible to ethically justify the need for the energy and societal transition.

6.2.3. *Individual or institutional responsibility?*

It seems accepted that the energy transition entails, at least in part, a certain sense of responsibility[6]. The notion of responsibility serves both to identify the possible actors of this transition, but also to justify its necessity. This dual status finally brings us to a more fundamental question: who is subject to this responsibility? Who is responsible for environmental damage? Answering this question is not insignificant, because it would make it possible to identify ways to ensure the effectiveness of the energy transition while giving full meaning to its societal dimension.

Assigning responsibility for ongoing climate change is perhaps less straightforward than it seems. It is true that the idea that individuals, at least those in industrialized societies, who act as citizens and/or consumers, have a share of responsibility which is particularly widespread, in particular because we know the impact of different actions on the environment. This responsibility has varying levels of extensiveness, but some authors defend the global responsibility of the individual:

---

5 This also refers to the Lockean clause which, according to Nozick, prescribes that any appropriation of a common good is only fair if its potential users receive compensation for being deprived of it.

6 Regarding the different meanings of the term "responsibility", see Ogien [OGI 04].

> Everyday choices and acts by individuals play an important role for the future of political, social, and economic life. In short, every person is part of global responsibility-taking [MIC 03].

Nevertheless, a first remark is somewhat necessary: can we really uniformly speak of individuals while obscuring the social, political and economic contexts in which they operate? It is also worth considering the very concept of responsibility. First, we can see that the responsibility we are reflecting on is of a *causal* nature. Besides the fact that this is only one type of responsibility [OGI 04], it seems to lead to a binary distinction: either I am entirely responsible for ecological problems, or I have no responsibility. To us, things seem more nuanced. Thus, and in the context of the question that concerns us, it seems relevant to us to distinguish *backward-looking* (retrospective) and *forward-looking* (prospective) responsibility[7]. This distinction makes it possible to propose several ideas: (1) individuals may not have to assume retrospective responsibility, in particular when they did not have reasonable alternatives; (2) conversely, individuals with reasonable alternatives can be assigned *prospective* responsibility; (3) there is a forward-looking responsibility of institutions (governments or companies) whose substantial approach is as follows: they must provide individuals with alternatives that allow them to choose the option with the weakest ecological impact.

To what extent can and should we hold individuals accountable for climate change and therefore impose on them the obligation to actively participate in the energy and societal transition? Let us take two simple examples[8] to discuss this point. Marie lives in a society where public transport is particularly developed, which is not the case for Agnès. In other words, Marie's decision-making context allows her to choose public transport and thus reduce her impact on the environment. Conversely, Agnès' context makes it difficult or more complicated, more costly (and not only in the economic sense of the term) to make the same type of decision. The alternatives offered to them are therefore radically different. Consequently, asking them both to use public transport does not have the same weight or the same ethical force: for Marie, the requirement seems at least reasonable when it is perhaps (probably) too demanding in the case of Agnès. This does not preclude the fact that, in both cases, we consider it ethically responsible to use public transport.

Now imagine that Anne and Sylvie would both like to buy a house. Anne lives in a society where several craftsmen propose building houses with low energy consumption both in their construction and in their use. Sylvie lives in a society where such craftsmen are rare and 40% more expensive. It appears here that the ethical requirement to build an energy-efficient house has a greater moral and financial cost for Sylvie than for Anne, regardless of their positions. Indeed,

7 The rest of our text is based on the positions defended in Fahlquist [FAH 09].

8 The first case takes up and adapts the example of Fahlquist [FAH 09].

regardless of Anne's position on this issue, it is not difficult for her to have such a house built. *On the other hand*, whatever Sylvie's position, *especially* if she wants to build an ecological house, taking such a decision has a significant cost.

What do these examples show? First, the degree of responsibility of Marie and Agnès in the first example and the degree of Anne and Sylvie in the second are not equivalent. Next, this degree depends on the context in which they must make their decisions and therefore take action. Finally, this context and the social infrastructure in which the decisions are made depend on political and institutional decisions. Therefore, it is fundamental to recognize the importance of the alternatives offered to individuals when it comes to holding them accountable to certain actions. In other words, attributing responsibility to individuals cannot be thought of in a uniform and unilateral way. Moreover, responsibility must be assessed in the light of the institutional context in which they operate. The structural and societal obstacles that can prevent individuals from making decisions and therefore from acting in favor of the environment must be taken into consideration. They form an integral part of any ethical reflection on the energy and societal transition, the latter term taking on its full meaning here.

What exactly do we mean here by the decision-making context? Or more precisely what is the important element that we must take into consideration? It is a question of defining what the reasonable alternatives are which will make it possible to define an individual's responsibility. In the case that individuals are offered reasonable alternatives in favor of ecology, they can be held responsible for their actions. Conversely, if this is not the case, they may not feel a sense of responsibility engaged. *Four factors make it possible to define reasonable alternatives which are at stake here* [FAH 09]. The first factor is the cost: it is not possible to have the same demands on two people with different incomes if the cost of the action is high. We must add the requirement that "good" options, or *valid options, are offered*: if I want to buy local food to reduce the carbon footprint of my diet, but the stores around me do not sell any, the fact that cost is not an issue is not a sufficient criterion. Next, one must consider the *substantial inconvenience of this action*: I cannot really be held responsible for going by car to my workplace if it is dangerous to take a bike because there is no cycle path to use. Finally, it is necessary that *information becomes available so that each individual can understand the environmental impact of their actions*.

Consequently, individuals cannot be held to have a backward-looking responsibility if they did not have reasonable alternatives, but it is nevertheless possible, in certain cases, to assign a forward-looking responsibility to them, which we will now define.

The idea of forward-looking responsibility ultimately comes down to saying that when an individual has the possibility to act in such a way so as to protect the environment, reduce their carbon impact, etc., they have an ethical (or moral) obligation to accomplish such objectives. But, and as we have said before, an individual who does not have this possibility cannot be held responsible for this impossibility and is therefore not subject to the same obligation. As we have also seen, these possibilities depend on the institutional, social and economic context in which individuals find themselves. Political institutions and economic actors (such as multinationals) have resources and means which are far superior to those of individuals, and it is therefore possible for them to open up the choice of possible actions. For them, it is not just a possibility, but the very idea of corporate social (or societal) responsibility creates an obligation on the part of companies. States on the other hand must allow individuals to choose between different possibilities so as to perform the action that seems best or right to them.

Therefore, political and economic institutions are responsible because they have the means and the power to offer people reasonable alternatives. We can now better understand how *individual and institutional responsibilities are intrinsically linked. The more institutions offer reasonable alternatives, that is, assume their responsibility, the more individuals can be held responsible if they do not perform ethical actions, in this context those in favor of environmental protection*. Applied to our purpose, this means that the energy and societal transition can only be accomplished to the extent that institutions allow individuals to choose actions that are ethically in accordance with the requirements of such a transition. *It is only when institutions present themselves as responsible that it will be possible to expect responsible behavior from individuals*. As it is the responsibility of institutions to enable individuals to be effectively responsible, *it is up to them to ensure that they have adequate institutions* [SHU 88]. In other words, *the duties are bilateral and reciprocal*, even if they operate on different levels. It is only by combining these responsibilities and obligations that the energy and societal transition can be effective.

We now have a good grasp of the dynamics of the relationship between individual and institutional responsibility. Now that the issue has been ethically grounded, the question remains surrounding its effectiveness and applicability. For example, how do we ensure that individuals act in accordance with their values? What are the means available to institutions, or States, to encourage individuals to act in favor of the energy transition?

6.3. From theory to action: the question of influence

6.3.1. *Individual and societal values: which will act in favor of the energy transition?*

We have seen that the ethical requirement could have different degrees depending on the options that are offered to individuals, but we have not yet addressed the question of whether individuals consider it to be fair on them, according to their system of values and preferences, to participate in the energy and societal transition. We somehow presupposed that it was good to act in favor of the energy and societal transition because it was a good and ethically justified action to protect the environment and/or fight against climate change, but we neglected to ask ourselves if this reasoning was shared by the individuals concerned, and if it was, if these same individuals have acted accordingly. In this way, several works show that *there is a gap between the conceptions we have (or think we have) and the actions we carry out*. The gap between the ethical conception that we declare to adopt in such a situation and the action accomplished is studied by Bazerman and Tenbrunsel in their book *Blind Spots* [BAZ 13]. This book is not about *what we would do*, what we say we would do, if we found ourselves in such a situation and what ethical behavior we would then adopt, as opposed to what we actually do (?) *when faced with an ethical dilemma*. Central to their analysis is the impact of cognitive biases on decision-making processes and what they call behavioral ethics. In other words, we are in favor of change regarding the measures to be taken to avoid climate deterioration, and we believe that it would be ethical to adopt such a behavior, but we do not act in accordance with this thought pattern. But maybe it is possible to get people to behave in accordance with their values and preference systems?

Education is probably the most commonly used method. It is supposed to allow individuals to have the information necessary for their decision-making. But to what end? It would seem that educating by informing does not give us the expected results. Thus, we have been able to observe that the consumption habits of McDonald's customers have absolutely not changed, despite the caloric analysis of food being displayed and visible. While New York City has implemented such a display, one study concludes:

> Over time, customers started to ignore the labels. More significantly, at no time did the labels lead to a reduction in the calories of what diners ordered. Even if people noticed the calorie counts, they did not change their behavior [CAR 15].

This attempt therefore ended in failure. However, other incentives could be used. A well-known example is increasing price of cigarette packs in the real or perceived hope to reduce consumption. This has effectively had the effect of reducing the number of smokers, but not of eradicating smoking altogether. People young and old, who are fully aware of the dangers associated with smoking, continue to smoke despite this price increase. We can therefore legitimately think that these two types of incentives, educational and economic, are not sufficient and have no real impact on the behavior of individuals. Hence, there is need to turn to other solutions [CON 16, SIN 15].

Nudges, these now famous soft influences, seem to be a way to correct "bad" decisions, that is, irrational decisions that are not in accordance with the preferences declared by individuals. They can influence the decision-making process, even if they focus on preferences and not ethical conceptions[9]. Their apparent simplicity, even their obviousness, nevertheless hides a fundamental question: is it possible, from the influence on a decision, to change the overall pattern of decisions that concern a specific problem, that is to say, the behavior to be adopted in favor of energy and societal transition? Can nudges allow for lasting change in the behavior of individuals or are they reduced to one-off attempts whose success is not even guaranteed?

6.3.2. *Developing the use of nudges in public policies*[10]

Efficiency is a recurring argument which is constantly put forward by the supporters of nudges. The latter would be more effective because of their low-cost implementation and their impact on the behavior of individuals. It is appropriate here to rely on a few examples in order to better understand what is at stake. The best-known example is probably that of the fly in the urinal (Figure 6.1). This nudge was implemented at Schiphol Airport (Amsterdam), which wanted to both improve the well-being of users while reducing expenses related to cleaning toilets. The nudge, introduced in 1999, consists of placing an image of a fly in the urinal which the men cannot help but aim at. The figures put forward indicate a reduction in cleaning expenses of around 80%.

Another particularly interesting example is the strategy implemented by Opower in California to reduce energy consumption. Indeed, while 90% of Californians consider energy issues to be "very" or "extremely" important and 98% of them say they are trying to reduce their energy consumption, it is clear that this does not seem to be the case. We see this particularly during the power cut that occurred during the 2001 heat wave[11]. Based on the conformism bias[12], which indicates that individuals will tend to conform to the social norms of the group to which they belong, Opower developed an energy report for each household. Figure 6.2 shows that this is based on comparing the household in question with 100 households closest to it, associated with a small drawing that indicates the quality of the behavior of the household concerned. The result obtained is a drop in electricity consumption of around 2% per household.

---

9 Even though our preferences may have ethical foundations, evaluating this issue would lead us to discussions that are well beyond the scope of this chapter.

10 We resume here the elements present in our article [BOZ 18].

11 According to an Opower survey carried out by a working group led by Robert Cialdini.

12 Highlighted in particular by Asch [ASC 51].

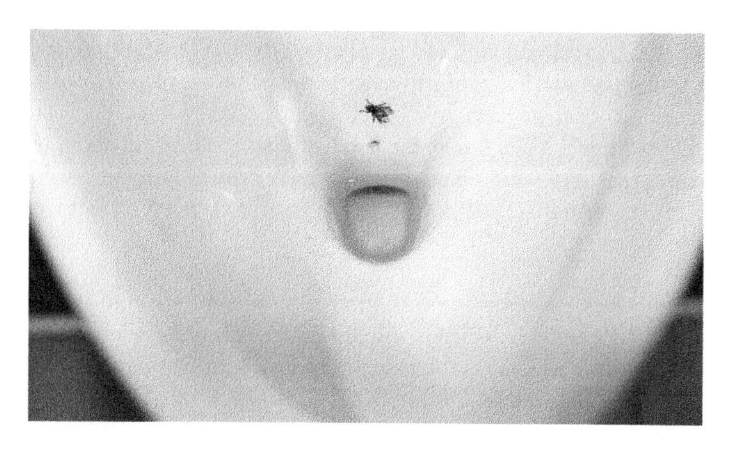

Figure 6.1. *The fly in the urinal (source: gettyimages.com)*

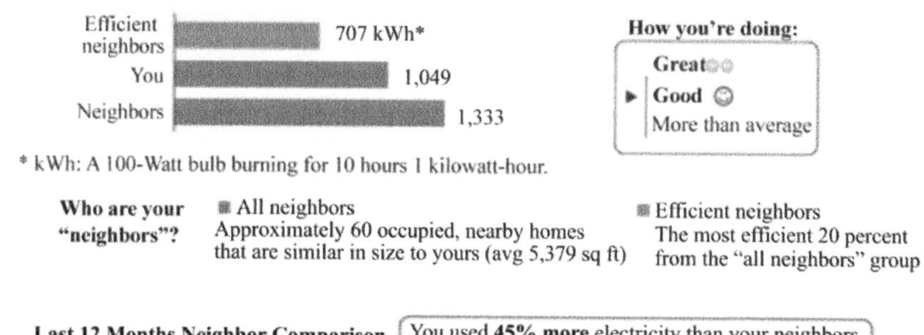

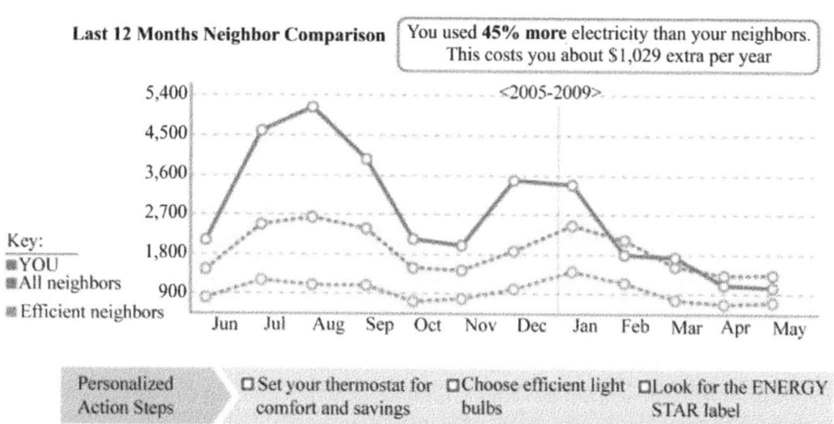

Figure 6.2. *Energy report from Opower (source: Opower's Home Energy Reports by Opower). For a color version of this figure, see www.iste.co.uk/robyns/smartusers.zip*

The two examples cited here are characteristic of how nudges are used: it is a question of finding a "soft" way of influencing the behavior of individuals. By soft, we mean "without legal or economic sanction". In other words, there is criticism surrounding "hard law"[13]. The law in the conventional sense of the term, understood as a set of rules that are based on sanctions in the event of violation, would not be sufficiently effective and would not or no longer make it possible to ensure the behavior of individuals.

It therefore seems necessary to add another system of tools to the mix which would make influencing behavior possible. From a conceptual point of view, it is a question of thinking about new forms or new levels of normativity. Those who defend nudges, with Sunstein at the forefront, do not claim that only nudges should be used, rather that they must be thought about as complementary and integrated into a set of available tools which are in the hands of the legislator. They therefore wish to encourage specific behaviors on the part of individuals, who are subject to their own will[14].

Nevertheless, we can highlight the fact that if nudges are not the only tools used, the experimental methodology on which nudges are based has an increasingly important place in the development of public policies.

The lesson that we can draw from the use of nudges is that in order to promote certain gestures, certain behaviors or certain policies, it is necessary that the ethical questions linked to the energy transition have a response that is sufficiently elaborated so that we can justify a certain number of axiological conceptions. In other words, as we have seen, it is a question of being able to produce arguments that are in favor of the transition. However, we can observe that this would call the neutrality toward conceptions of good which are generally attached to modern democratic states[15] into question. One way to counter this could be to substitute a conception of justice for a conception of good, but then would this really fit in with the idea of nudges as they have been initially conceived? Thus, it would be necessary to justify imposing such a conception (whether it is a fair or good conception). In other words, the question arises around justifying paternalistic interventions[16] on the part of a state. These questions open the door to the issue of

---

13 It would be tempting here to tilt the nudges on the side of soft law, but this is a step that we will refrain from taking, as it would indeed require an intentional study.

14 We are fully aware here that we are using a language close to the first legal positivists, notably Bentham and Austin.

15 See Rawls [RAW 09] on this point.

16 Whatever the form of paternalism. For a classification and analysis of the different forms of paternalism, see Le Grand [LEG 15].

nudges as measures that relate to a form of paternalism of means and not of ends [LEG 09].

This therefore amounts to saying that for nudges to be used with regard to the issue of energy and societal transition, it would first be necessary to *establish the values that should be promoted, made consensual and shared* [GOD 15, HAY 16, ROS 16, SHU 14]. In section 6.2, we believe we have established the ethical justification for promoting the energy and societal transition and have emphasized the need to provide reasonable options to individuals. However, it remains to be affirmed that these conceptions are shared and understood. This question is important because here we are touching on the preference systems of individuals: by aiming to influence their actions, we cannot ignore the risk of penetrating the individual's intimate space, their global and internal preference systems. In other words, when a nudge influences decision-making, is it limited to momentarily changing the preference of the individual or does it change the overall preference system of this same individual? If some people may argue that this can be justified in any case so long as the promoted action is good and accepted as such, what would the position of these same people be if they thought that the action in question was not good? Thus, reflecting on nudges as a tool for influencing behavior in the context of public policies must go beyond the ad hoc nature of certain policies. Even when considering that only purpose can define the ethical or unethical character of a nudge, the problem of justifying the purposes and therefore the nudges themselves remains unsolved. However, it is clear that nudges very often operate by limiting their transparency and their process of transparency to a small group of experts who are in charge of developing these nudges. In other words, while nudges may seem attractive, they nevertheless raise a number of ethical questions. In order to understand them, it is now necessary to analyze the conceptual framework of nudges as well as their definition in order to understand their implications.

6.4. Nudges: ethical issues raised by the use of behavioral sciences to influence behavior

6.4.1. *Conceptual framework and definition*[17]

Sunstein and Thaler are relatively clear about the theoretical foundations of their work on nudges:

> But our basic source of information here is the emerging science of choice, consisting of careful research by social scientists over the past four decades. That research has raised serious questions about the

17 We take up the article by Bozzo-Rey [BOZ 16].

> rationality of many judgments and decisions that people make. [...] Hundreds of studies confirm that human forecasts are flawed and biased. Human decision making is not so great either [SUN 08].[18]

This can be grouped under the term behavioral sciences, and more specifically scientific approaches to behavior that identify the predictable irrationality of human decisions. Such sciences try to challenge the vision of classical economics which considers the individual as a *homo economicus* guided by their interests and making rational decisions. On the contrary, they consider that individuals often act irrationally, therefore against their own interests due to cognitive biases.

The theory of nudges is also articulated around two key concepts: the choice architecture and libertarian paternalism. Sunstein and Thaler thus underline the importance of the context in which decisions are made, which justifies the idea of modifying it to influence them. In addition, it is desirable to intervene in the decision-making process as late as possible in order to influence it; acting on the context seems to be adequate in this regard. It is thus necessary to more precisely formalize what is meant by "context" in the decision-making process. This is where the concept of "choice architecture" comes into play. The person who sets it up or modifies it is a "choice architect", a term which can therefore cover very different realities: from the public decision-maker to the parent. In other words, in any nudge, there is a "nudger" (the choice architect) and a "nudgee" (the person who is subject to the nudge); all of it takes place in a choice architecture. Thaler and Sunstein present nudges as shaping of the choice architecture that alters the behavior of individuals in a predictable way, without prohibiting or significantly modifying economic incentives. Therefore, the intervention must be easy and inexpensive to avoid; the low exit cost is ultimately an important argument for these two authors. A quick examination shows that this undoubtedly necessary condition is however not sufficient: *in our opinion, a choice architecture that merely informs individuals can modify their decisions, but cannot be considered as a nudge*. The idea underlying many nudges is indeed to exploit our cognitive biases. We must also emphasize that for Sunstein and Thaler, there is no "neutral" or avoidable choice architecture: any architecture will influence the individual in one direction or another and any decision is made in a particular context.

The second key concept developed by Sunstein and Thaler and that is supposed to provide a strong foundation for their theory is that of libertarian paternalism[19]. Much of the criticism of nudges has focused on the latter and either points out that it is neither paternalism nor libertarianism per se [HAU 10], or that it is just an

18 See also Cartwright [CAR 14]. For a more historical presentation, see Thaler [THA 15]. For a presentation of advances in behavioral economics, see Camerer et al. [CAM 04].

19 Sunstein also talks of soft paternalism in Sunstein [SUN 15].

oxymoron [MIT 05]. The issue is therefore at the level of autonomy and freedom, which we will come back to.

First, let us remind ourselves what is meant by the term paternalism. According to Gerald Dworkin,

> *X acts paternalistically towards Y by doing (omitting) Z*:
>
> 1) *Z* (or its omission) interferes with the liberty or autonomy of *Y*;
>
> 2) *X* does so without the consent of *Y*;
>
> 3) *X* does so only because *X* believes *Z* will improve the welfare of *Y* (where this includes preventing his welfare from diminishing), or in some way promote the interests, values, or good of *Y* [DWO 14].

Sunstein and Thaler consider a policy to be paternalistic

> if it attempts to influence the choices of affected parties in a way that will make choosers better off [SUN 03].

We will accept here Dworkin's definition (but there are variants and a certain number of subtleties as Le Grand and New [LEG 15] point out), which makes it possible to account for the different forms of paternalism and to understand that, even assuming it to be libertarian, Sunstein and Thaler's paternalism is likely to interfere with Y's freedom and/or autonomy. Libertarian paternalism is as follows:

> an approach that preserves freedom of choice but that authorizes both private and public institutions to steer people in directions that will promote their welfare [SUN 15].

Here, we can note that Sunstein and Thaler try to distinguish themselves from conventional paternalism by insisting that freedom of choice of individuals be respected. Without developing this point too much, we cannot neglect to emphasize that their definition of freedom is simplistic, since it considers that for an individual to be free, it is "enough" that the number of choices at their disposal is not reduced. Let us take an example: a conventional paternalistic measure which makes reducing smoking possible would be to simply ban cigarettes (we can then speak of coercive paternalism, like Sarah Conly [CON 15]), or, in a more moderate way, to impose taxes on tobacco. On the contrary, a libertarian paternalistic measure would be to impose repulsive information or images on cigarette packets, such as those which are now compulsory in the European Union (and in a number of other countries). Here, the context in which the decision to smoke is made has changed, but this has come without a ban or change in economic incentives. On the other hand, a form of "mental tax" is imposed on the consumer, by placing unpleasant messages or images in their view that will influence them. From this point of view, and if one follows the definition of freedom from Sunstein and Thaler, one interferes with autonomy rather than freedom.

We can now attempt to characterize the relationship between nudge and libertarian paternalism as follows: a nudge implemented without the consent of the individual and aimed at improving their well-being is an example of libertarian paternalism.

According to Sunstein and Thaler, one of the great advantages if not the main advantage of libertarian paternalism is that it does not involve coercion: it is always possible to make a decision that goes against a nudge. This possibility is fundamental. It implies that, from a theoretical point of view, it is possible to think of forms of paternalism that do not resort to coercion.

We are now able to give a clearer and more complete definition of a nudge, although, as we will see and since Sunstein and Thaler do not deliver a "canonical" definition, if not a minimalist one[20], the definitions may vary. In fact, the question no longer becomes a definition as much as what specifically is able to characterize an action as a nudge, the very characteristics that will enable one to identify a nudge. In other words, it comes down to what we can consider a nudge to be:

> To count as a mere nudge, the intervention must be easy and cheap to avoid. Nudges are not mandates. Putting the fruit at eye level counts as a nudge. Banning junk food does not [SUN 08].

Therefore, it is possible to come to the following definition of nudges:

> A nudge is a function of (I) any attempt at influencing people's judgment, choice or behaviour in a predictable way (1) that is made possible because of cognitive boundaries, biases, routines and habits in individual and social decision-making posing barriers for people to perform rationally in their own declared self-interests and which (2) works by making use of those boundaries, biases, routines, and habits as integral parts of such attempts [HAN 16].

Therefore, a nudge aims to change the behavior of individuals in a predictable way by influencing their judgment or choice. To do this, it depends on the presence of cognitive biases that are inherent in the decision-making process which prevent

---

20 "Any aspect of choice architecture that alters the behavior of individuals in a predictable direction without forbidding an option or significantly changing their economic incentives" [SUN 12].

individuals from performing the rational action that they should perform if they were really acting in their own interest. We must therefore remember that cognitive biases are an integral part of a nudge.

We can complete these definitions by thinking about the nudge as an alteration of the choice architecture which fulfils the following conditions: (1) it aims to modify the behavior of individuals via an interaction with their non-deliberative faculties (either by triggering them or by blocking them); (2) it aims to contribute to the well-being of the individual in question, or to a common good (such as the well-being of society as a whole, or for us the protection of the environment); (3) it does not impose a ban and does not change economic incentives in any significant way (i.e. the intervention must be easily avoidable, at a low cost).

6.4.2. *Elements of a critique of nudges*

As seductive as nudges seem to be, they do raise a number of questions that their opponents have been quick to point out. The first criticism is ultimately a consequence of the presuppositions of the very theory of nudges. As we have seen, Sunstein and Thaler claim that they use findings in behavioral science as the epistemological foundation of their theory. These discoveries insist on the place of irrationality in human decisions while relying on an implicit model of rationality. Moreover, it results in distinguishing *econs* from *humans*. Sunstein and Thaler thus establish such a distinction:

> If you look at economics textbooks, you will learn that homo economicus can think like Albert Einstein, store as much memory as IBM's Big Blue, and exercise the willpower of Mahatma Gandhi. Really. But the folks that we know are not like that. Real people have trouble with long division if they don't have a calculator, sometimes forget their spouse's birthday, and have a hangover on New Year's Day. They are not homo economicus; they are homo sapiens. To keep our Latin usage to a minimum we will hereafter refer to these imaginary and real species as Econs and Humans [SUN 08].

In other words, "econs" are perfect calculators as imagined in the theories and models of classical economics, while "humans" are ordinary people who often tend to make mistakes and do not make the best decisions. The question of the individual's vision proposed by nudges therefore arises: are we irrational individuals whose irrationality should, precisely, be corrected? This is the first part of the criticism: nudges do not consider us to be rational individuals. The second part is to say that they do not even consider us as autonomous individuals, capable of self-determination. These two parts are intrinsically linked:

rationality enables autonomy and autonomy builds on rationality. To attack one is like attacking the other. Andreas Schmidt [SCH 19] defends the idea that nudges will in many cases improve our ability to act rationally. It is based on an "ecological" conception of rationality [MOR 11], which affirms that an individual's decision-making process is rational in a specific environment insofar as, taking into account their capacities and psychological tendencies, they allow them to achieve their ends, in this given environment. The same logic would apply for self-determination. Rozeboom opposes such an interpretation of nudges [ROZ 20]. He considers that appealing to or improving the rational capacities of an individual is not a sufficient argument to consider that they are treated as a rational individual. For example, if individual *A* lies to individual *B*, they directly mobilize the rational capacities of *B*, but do not necessarily consider them a rational individual. On the other hand, what argument could be mobilized to defend the idea that individuals who behave irrationally necessarily want the help of a nudge to act rationally? In other words, and in all cases, nudges would then be treating individuals as irrational beings. For Rozeboom, the ethical condition required to use nudges is precisely to consider agents as rational beings, which implies wanting to preserve and recognize the intrinsic value of an individual's capacity to act.

To continue this first criticism, Hausman and Welch [HAU 10] focus their attention on the links between nudges and deliberative capacities, the latter integrating a practical reasoning at the end of which a decision is made. These authors analyze nudges as instances of paternalism because "in addition to or apart from rational persuasion, they may "push" individuals to make one choice rather than another" [HAU 10][21]. Indeed, freedom in the sense of freedom of choice is not restricted, but from the moment a nudge does not take the form of rational persuasion, an individual's autonomy is not fully respected. Indeed, their decision, and therefore their actions, will then not be the expression of their own evaluations and deliberations, but rather the result of the strategy that is put in place by the choice architect. The latter shapes the choice of individuals. In other words, nudges represent a threat in the sense that they interfere with or even diminish the exercise of the deliberative capacities of individuals.

However, some authors go further, since they consider that this way of shaping decisions strongly interferes with the freedom of individuals and amounts to manipulation. Thus, Grüne-Yanoff [GRÜ 12] considers that the justification that is advanced by proponents of nudges by defending a libertarian paternalism to limit freedom of individuals on behalf of their well-being is not conclusive (where freedom is not reduced to freedom of choice, but rather thought of at a minimum as non-interference). Indeed, the conception of well-being respects neither subjectivity nor the plurality of individual values. Interference with freedom implied

21 On this point, also see [SUN 06].

by the use of nudges is therefore not justified. Wilkinson [WIL 13] considers that such interference is in fact manipulation. It can be characterized by three elements: (1) a perversion of the decision-making process [RAZ 86]; (2) the intention on the part of the person who is manipulating; and (3) violation of individual autonomy. Wilkinson is nevertheless nuanced, since he indicates on the one hand that all nudges do not necessarily relate to manipulation (this would be the case, e.g., of nudges which do not intentionally target individuals to be manipulated), and on the other hand, manipulation can be consented to and as such is acceptable[22].

We can therefore clearly see to what extent the various criticisms of nudges bring into play the questions of autonomy and freedom. A final criticism, made by White [WHI 13], is also interesting, because it no longer focuses on the "nudgee", but on the "nudger", on the choice architect. White ultimately invites us to draw conclusions from the justification for using nudges proposed by Sunstein and Thaler via behavioral sciences. Since we do not make good choices, it is justified that other people help us to make better ones. In this case, public decision-makers, or "nudgers", help us. But then an epistemic question arises: how could a "nudger" know the preferences of individuals, assuming that the latter are able to know them themselves? White answers two aspects. First of all, he considers that each individual is the best judge of their own interests, and then, the consequence of this affirmation is that if a public decision-maker makes a judgment on the choices of other individuals, this can only be done based on his own criteria. In other words, nudges can only be the expression and imposition of the value system of a "nudger" on one or more "nudgee". As such, resorting to nudges cannot be justified.

Indeed, we could add an element that questions the conception of the individual conveyed by the nudges not going *upstream* but rather *downstream*. What type of individual would result from an intensive and systematic use of nudges? Would it not be better to try to allow individuals to emerge who are aware of their choices, of their possible mistakes and of what motivates them? By reducing the decision-making capacity of individuals and the frequency with which they use their deliberative capacities, are we not undermining the very foundation of democracy, which is based on the capacity of individuals to make free and enlightened decisions?

6.4.3. *What place is there for nudges in the energy and societal transition?*

In light of these different criticisms and the different elements that we have mentioned, what place should be given to nudges in the energy and societal transition?

22 Of course, these points deserve analysis and discussion, but this is not the subject of this chapter.

A first remark consists of questioning the globalizing unification that nudges seem to put into effect. Everything occurs as if all behaviors and therefore the changes that come from them were equivalent. In this case, can we really consider that a change in behavior concerning economic decisions would be equivalent and would obey the same laws as a change in behavior concerning ecology? We can also put it in another way: admittedly probably in a somewhat caricatural way, is aiming correctly when using a urinal equivalent to ensuring the well-being of future generations through the protection of the environment? Are we not faced with questions that not only necessarily directly benefit the individuals who will act in a given way, and therefore do not in fact correspond to their preferences, but rather relate to society as a whole? In other words, is it not a question of asking whether the nudges make it possible to adequately address the question of common people, of issues that pertain to the common good?

The second point lies in the fact that any nudge would only be able to ensure an extremely partial change in behavior, but would not generally intend to promote a social conception or a lifestyle oriented toward sustainable development. Indeed, if this were the case, nudges or more precisely libertarian paternalism would be a paternalism of ends and not of means [LEG 15]. Only particular actions in specific choice architectures are concerned. Nudges thus completely set aside the question of the intrinsic and lasting motivation of the individual to behave in a coherent manner, even during a change in the choice architecture [BOV 09]. *Nudges therefore evade reflections about the values that could guide individuals in a more sustainable way.* Therefore, nudges can only have short-term effects and only reinforce a certain type of behavior, even if they minimize it [THØ 09].

Nudges also raise political questions, since they involve changing behavior in an imperceptible way[23]. More specifically concerning the energy and societal transition, are nudges really able to respond to the challenge they represent and whose response requires a radical change in certain lifestyles, infrastructures and institutions? However, this must be based on a social and political dynamic that cannot be achieved without public debate. In other words, because they do not figure in public debate due to the very fact that they act on the non-deliberative capacities of individuals (even if they take the more scientific and apparently axiologically neutral term of cognitive bias), would nudges not constitute within a highly paradoxical movement an additional means that *favors the dominant paradigm of a deregulated economy based on growth? Do the nudges have the capacity for change that is necessary and adapted to energy and climate issues?* Of course, it would

23 And not only invisible, since transparency is one of the conditions for the democratic exercise of nudges [SUN 16].

always be possible to conclude that nudges can play a part in behavioral changes, but that they are not the only solution, nor are they the miracle solution. And indeed, we have seen that even their most ardent defenders do not consider that public policies should be limited to the use of nudges. However, we can only highlight the extent to which public and private decision-makers seem fascinated by the low cost and the apparent scientific nature of these tools. Perhaps, it should be emphasized that these are only tools to influence behavior among others. It is therefore likely that it is not only nudges that must be questioned, but the place now occupied by behavioral sciences in the development of public policies.

We thus come to a series of elements that are necessary to keep in mind. First, it is accepted that nudges have a low rate of effectiveness: around 10%–15% [ARN 16]. Second, nudges are only one method among others. They should not replace, for example, education. Rather they should be complementary, but not exclusive. Third, if it is decided that such tools should be used, the risk emerges to entrust their development to nudge experts who would replace the democratic decision-making process and impose a value system, a purpose, without leaving room for debate. Finally and consequently, transparency and public consultation are necessary and essential elements that must integrate the process of developing any policy based on the use of nudges. If the energy and societal transition was to use nudges, this could only be considered as a complementary approach and should in no way replace the organization of a public debate which allows for awareness and individual reflection on current lifestyles. Finally, in principle, nudges focus on individual decision-making processes, but this should not mean that we neglect to reflect on the institutional responsibility of the alternatives offered or not offered to the individuals, as well as the distribution of tasks that allow this transition to be a success.

6.5. Conclusion: the necessary political dimension of ethics

The energy and societal transition poses important and complex problems, both theoretical and practical in nature. We have seen that relying solely on an individual moral obligation cannot be justified because it in fact establishes an inequity between individuals and sets aside two elements: the place of organizations and the role of social and political institutions. As a result, and to use the expression of John Broome [BRO 12], the energy and societal transition should rather be considered as a paradigmatic example of the *need to unify private (individual) and public (institutional) ethics*. From this point of view, nudges could seem to be an adequate response: public decision-makers put in place public policies that are based on the interests of individuals "as judged by themselves". However, the potential risk could be illegitimate and unjustified interference with the autonomy and freedom of the individuals concerned. Additional to this could be a prevention of a responsible and autonomous individual to emerge, which enables the proper functioning of democratic societies. Consequently, the energy and societal transition should rather be understood as the expression of the current challenge of our democracies *to combine ethics and politics with respect for individuals, protecting their interests as well as those of future generations.*

Postface

The authors of this book, two engineers, a sociologist and a philosopher, take us on a great journey through the energy transition and its societal consequences. Upon reading it, one can feel optimistic when witnessing the sum of technological, economic and social ingenuity of which mankind is capable, while also wondering whether local and global solidarity, which appears as essential to succeed in this transition, will actually come to pass. To succeed in the energy transition, it is necessary to collectively recognize the problems, climatic in nature, for example, which have not yet been resolved in our society, since climatosceptics remain relatively numerous (20% of French people are skeptical according to a recent survey). There is therefore a lot of work to be done to raise awareness, which this book allows for, particularly as it does not appeal to catastrophism, but rather is oriented towards a search for concrete actions.

The authors show in a relevant fashion that in order to experience the energy transition, it is important to work at several levels. Thus, climate change which is caused by greenhouse gases is linked not only to the energy choices made by each nation, to access to raw materials, to the development of cutting-edge technologies that combat CO_2 emissions, but also to the type of economy, ways of life and consumption patterns, as well as many other factors such as the standard of living within populations and corresponding town planning. Here, we can speak of "transitions" in a complex world where interactions are multiple and of different orders.

In this regard, let us highlight the European initiative of the Green Pact which proposes a new growth strategy for the European Union with a view to creating a more "sustainable" and fairer society, integrating environmental, energy, health and food issues, while taking into account social, economic, financial, ethical and political issues. The fundamental systemic dimension of interdependencies is necessary, but often difficult to integrate into the operational management of complex processes that involve multifactorial issues. This jeopardizes their coherent and effective implementation. Yet, it is this challenge that the Green Pact wants to tackle. Within this vision, nature, matter, living species and territories are no longer primarily resources to be exploited by a human "master and possessor of nature". This becomes a real lever to be able to change our modes of production, consumption and therefore of living.

The great advantage of this work is to be able to disentangle many aspects of the complexity inherent in the situation, to never use a single key parameter to solve all problems, such as the "all technological" parameter, which is nonetheless still often advocated for. The authors underline the links between energy issues and lifestyles, particularly with regard to the energy consumed for travel, heating or food, without forgetting the question of the buildings in which we live. "So would there be no smart buildings and smart cities without smart users?" This requires us to adjust policies in terms of training and education, so that they are firmly rooted in sociological analyses that are linked to acceptance and appropriation by users.

One of the merits of this work is also to propose possible solutions within the framework of the territories and the university, in an intelligent alliance between different organizations, each holding a share of power to move things forward. Different examples of actions within universities in their territory are offered: they explain the particular interest of these alliances. Thus, in their diversity of training for engineers, managers, economists, specialists in health and the human and social sciences, when higher education establishments like the Université Catholique de Lille combine their knowledge and know-how, in close collaboration with the city, like the metropolis of Lille and the Hauts-de-France region, concrete paths for a "sustainable future" in the corresponding territory may emerge. In this sense, the *living labs* of this same university, such as Live TREE which seeks to promote the UN's sustainable development objectives to implement actions for the energy transition, represent real third-party collaborative spaces, sources of collective intelligence that our society is in need of. The demonstrators that these third parties produce make the search for sustainable solutions to experience the energy transition relevant.

To meet the challenges analyzed in this book and to act according to the avenues proposed, the questions of training and education are major. With regard to the university, which is presented here as a hub for working on the technical and societal changes necessary for the energy transition, training in ethics from within the programs undertaken by engineers, economists, lawyers, political scientists or managers seems essential. The chapter about the ethics of energy and societal transition emphasizes the need to take future generations into account in today's projects, as the philosopher Hans Jonas already said 40 years ago through his *responsibility principle* and the categorical imperative that followed: act in such a way that the effects of your actions are compatible with the permanence of an authentically human life on earth. How do ecological considerations for "taking care of our common home" come to transform our ethical reasoning, at the individual level, at the level of a company and a community, and at the level of national and international politics? How and to what extent can the living world inspire mankind in their modes of organization? What are the consequences of technical, economic and social choices?

We can no longer speak of "sectoral" ethics like we used to, since "everything is linked". Training in ethics, "from within the apprenticeship of professions", must take concrete behavior and the necessary changes in lifestyles in society into account. Professions themselves should be modified, at least the way of exercising them and therefore of training for them, with a broader sense of responsibility in the service of the common good, and not only within the ethical framework of a profession. In the ethics of desirable futures, key elements such as sobriety are combined with the most advanced technical solutions, which upsets our habits in terms of the notion of progress. The living conditions of future generations may be harsher than ours today, particularly in terms of the climate: for them, we must realize that today's technological innovation can only be seen as real progress if this innovation is truly "sustainable".

Many authors thus believe that the ecological crisis is an illustration of the death of a progressive paradigm that has had its day. In this context, the words of Pope Francis in his encyclical *Laudato si'* [FRA 15] resonate more strongly than ever: "It is not enough to reconcile the protection of nature and financial profit in a happy medium, or the environmental preservation and progress… It's about redefining progress. Technological and economic development that does not leave a better world and an integrally higher quality of life cannot be considered progress" (paragraph LS 194 in [FRA 15]). This progress is not to be confused with economic growth, with an increase in technological power, the accumulation of material wealth and an increase in GDP without neglecting these factors. "There are not two separate crises, one environmental and the other social, but a single and complex socio-environmental crisis" (paragraph L1 139 in [FRA 15]).

In the background of the work in ethics, questions of anthropology arise in a renewed way. It is neither more nor less our ways of seeing the world and "of being in the world" which determine the technical and economic choices to make so we can take care of the common planet or choose not to, of biodiversity, of people in the diversity of their geographical, cultural circumstances and climate-related migrations. This is why we speak of an "integral ecology", which is based on the relationship of human beings with their environment, integrating techno-economic development, social relationships, cultural and spiritual values and finally quality of daily life. The imbalance between these developments, relationships and values can be recognized as the anthropological root of the ecological crisis. Training in energy transitions thus goes hand in hand with openness to the spirit of integral ecology, to the humanism that such a vision is in the process of bringing out. We can already see it in the world of young people who are passionate about taking up the ecological challenge, despite slow reactions of major industrial countries. For example, a Climate Convention organized in the fall of 2022 at the Université Catholique de Lille will stimulate the entire institution: 100 students and 50 staff members will, with the support of many experts, work to propose possible decisions to be implemented so that the university can meet climate challenges.

In our universities, civic engagement has long been encouraged in the context of community life, and many student associations are thus invested in the ecological challenge. A new training pedagogy, named service learning, is beginning to take shape, this time at the heart of academic courses which train people for professions. The idea is to promote learning a profession by experimenting with a free service for people who are disadvantaged in the field in question. Thus, for example, undergraduate students in "sustainable development technology and management" from the Institut Catholique de Lille meet residents in the university district who have very poorly insulated housing. In dialogue with the inhabitants concerned and with experts in sustainable habitats, they offer technological solutions free of charge and possibilities of financial support in order to jointly meet the energy challenges of these inhabitants. The difficult question of the energy transition leads to new training pedagogies which should renew the interest of our students for their future professions.

In her book *Good Economics for Hard Times* ([BAN 20], in French), the 2019 Nobel Prize winner in Economics, Esther Duflo, recounts her experience as a member of the poverty reduction laboratory at the renowned Massachusetts Institute of Technology. With researchers from all over the world and members of NGOs, she shows that taking up the challenge of ecology and addressing the challenge of poverty and migration are one and the same. According to her, this opens up new intelligible horizons for a different world to be built, a world in which social friendship and fraternity could illuminate the paths of the future. "There is always something to do to change the world" is the leitmotif and driving force of this committed economist. Such testimonies suggest that the challenges to which this book makes a very useful contribution will be met as far as the humanist audacity of our engineers, researchers, economists and lawyers is able to extend. Particularly, if they do not forget to engage in essential dialogue with all inhabitants of the territories concerned, including the most modest. There is our hope.

Thierry MAGNIN
Deputy President-Rector for Humanities
and student life at UC Lille

References

[ADE 18] Ademe, Trajectoires d'évolution du mix électrique 2020-2060, Ademe, 2018.

[ADE 20] Ademe, Baromètre Ademe sur les représentations sociales du changement climatique, Ademe, July 2020.

[ADE 21] Ademe, Transition(s) 2050, choisir maintenant, agir pour le climat, Ademe, 2021.

[ADE 22] Ademe, Bilan GES organisation, available at: https://bilans-ges.ademe.fr/fr/accueil/contenu/index/page/bilan%2Bges%2Borganisation/siGras/1, 2022.

[ARI 04] Aristotle, *Politics: A Treatise on Government*, translated by William Ellis, available at: https://www.gutenberg.org/cache/epub/6762/pg6762-images.html, 2004.

[ARN 16] Arno A., Thomas S., "The efficacy of nudge theory strategies in influencing adult dietary behaviour: A systematic review and meta-analysis", *BMC Public Health*, vol. 16, no. 1, p. 676, 2016.

[ASC 51] Asch S.E., "Effects of group pressure upon the modification and distortion of judgments", *Organizational Influence Processes*, vol. 58, pp. 295–303, 1951.

[BAN 20] Banerjee A.V., Duflo E., *Economie utile pour les temps difficiles*, Le Seuil, 2020.

[BAR 14] Barre N., Roubaud M., *Les énergies renouvelables*, Éditions 10/18, Paris, 2014.

[BAR 20] Barrau A., *Le plus grand défi de l'histoire de l'humanité*, Michel Lafon, Neuilly-sur-Seine, 2020.

[BAZ 13] Bazerman M., Tenbrunsel A., *Blind Spots: Why We Fail to Do What's Right and What to Do About It*, Princeton University Press, Princeton, 2013.

[BES 15] Beslay C., Gournet R., "Les professionnels du bâtiment face aux enjeux de la performance énergétique : nouveaux savoirs, nouveaux enjeux", *Sociologies*, 2015.

[BOO 22] Bookchin M., *The Philosophy of Social Ecology: Essays on Dialectical Naturalism*, AK Press, Edinburgh, 2022.

[BOV 09] BOVENS L., "The ethics of nudge", in GRÜNE-YANOFF T., HANSSON S.O. (eds), *Preference Change*, Springer Netherlands, Dordrecht, 2009.

[BOZ 16] BOZZO-REY M., "Enjeux et défis de stratégies d'influence obliques des comportements : le cas de la législation indirecte et des nudges", *La revue Tocqueville*, vol. 37, no. 1, pp. 123–157, 2016.

[BOZ 18] BOZZO-REY M., "Les nudges face au changement climatique", in BAUDU A., SÉNÉCHAL J. (eds), *La conduite du changement climatique : entre contraintes et incitations*, LGDJ, Paris, 2018.

[BRE 02] BRENNAN A., LO N.Y.S., "Environmental ethics", in *The Stanford Encyclopedia of Philosophy*, Summer 2022 edition, ZALTA E.N. (ed.), available at: https://plato.stanford.edu/archives/sum2022/entries/ethics-environmental/, 2002.

[BRI 13] BRISEPIERRE G., "Les ménages français choisissent-ils réellement leur température de chauffage ? La norme des 19 °C en question", in *La sociologie de l'énergie 2 : pratiques et modes de vie*, BESLAY C. and ZELEM M.-C., Editions du CNRS, Paris, 2013.

[BRO 12] BROOME J., *Climate Matters: Ethics in a Warming World*, W.W. Norton & Company, New York, 2012.

[CAM 04] CAMERER C.F., LOEWENSTEIN G., RABIN M., *Advances in Behavioral Economics*, Princeton University Press, New York, 2004.

[CAN 18] CANEY S., "Justice and posterity", in KANBUR R., SHUE H. (eds), *Climate Justice: Integrating Economics and Philosophy*, Oxford University Press, Oxford, 2018.

[CAN 20] CANEY S., Climate justice, available at: https://plato.stanford.edu/archives/win2021/entries/justice-climate/, 2020.

[CAR 14] CARTWRIGHT E., *Behavioral Economics*, 2nd edition, Routledge, New York, 2014.

[CAR 15] CARROLL A.E., "The failure of calorie counts on menus", *The New York Times*, available at: http://www.nytimes.com/2015/12/01/upshot/more-menus-have-calorie-labeling-but-obesity-rate-remains-high.html?hp&action=click&pgtype=Homepage&clickSource=story-heading&module=second-column-region®ion=top-news&WT.nav=top-news&_r=0), 2015.

[CAS 20] CASSORET B., *Transition énergétique. Ces vérités qui dérangent*, De Boeck Supérieur, Louvain-la-Neuve/Paris, 2020.

[CER 13] CERTU (Direction générale de l'aménagement, du logement et de la nature), *Réduire l'impact environnemental des bâtiments, agir avec les occupants*, BEAUREGARD S. (ed.), CERTU, no. 275, 2013.

[COM 21] COMMISSION EUROPÉENNE, Mettre en œuvre le pacte vert pour l'Europe, available at: https://ec.europa.eu/info/strategy/priorities-2019-2024/european-green-deal/delivering-european-green-deal_fr, 2021.

[CON 15] CONLY S., *Contre l'autonomie : la méthode forte pour inspirer la bonne décision*, Hermann, Paris, 2015.

[CON 16] CONLY S., "When freedom of choice doesn't matter", *La revue Tocqueville*, vol. 37, no. 1, 2016.

[CRE 13] CREDOC, "Comment limiter l'effet rebond des politiques d'efficacité énergétique dans le logement ? L'importance des incitations comportementales", *La Note d'analyse*, no. 320, 2013.

[DES 95] DE SHALIT A., *Why Posterity Matters*, Routledge, London, 1995.

[DEV 16] DE VULPIAN A., *Éloge de la métamorphose. En marche vers une nouvelle humanité*, Saint-Simon, Paris, 2016.

[DEV 19] DE VULPIAN A., DUPOUX-COUTURIER I., *Homo Sapiens à l'heure de l'intelligence artificielle. La métamorphose humaniste*, Eyrolles, Paris, 2019.

[DUB 17] DUBRULLE L., "Quel sens donner à Live TREE", *Live TREE Mag*, no. 3, November 2017.

[DUR 20] DURILLON B., DAVIGNY A., KAZMIERZCAK S. et al., "Decentralized demand response considering residential profiles for load smoothing", *Sustainable Cities and Society*, vol. 63, pp. 1–11, 2020.

[DUR 21] DURILLON B., SALOMEZ F., DAVIGNY A. et al., "Consumers' sensitivities and preferences modelling and integration in a decentralized two levels energy supervisor", *Mathematics and Computers in Simulation*, vol. 183, pp. 142–157, 2021.

[DWO 14] DWORKIN G., "Paternalism", in *The Stanford Encyclopedia of Philosophy*, Fall 2020 edition, ZALTA E.N. (ed.), available at: https://plato.stanford.edu/archives/fall2020/entries/paternalism/, 2014.

[ECO 09] ECONOMIST IMPACT, The European Green City Index measures the environmental performance of 30 major European cities, available at: https://eiuperspectives.economist.com/sustainability/european-green-city-index, 2009.

[EKS 20] EKSTRÖM S., NOMBELA J.G., *Nous ne vivrons pas sur Mars, ni ailleurs*, Favre, Lausanne, 2020.

[FAH 09] FAHLQUIST J.N., "Moral responsibility for environmental problems – Individual or institutional?", *Journal of Agricultural and Environmental Ethics*, vol. 22, no. 2, pp. 109–124, 2009.

[FER 08] FERONE G., *2030, le krach écologique*, Grasset, Paris, 2008.

[FER 21] FERNBANK D., University of Reading Net Zero Carbon Plan 2021-2030, Report, University of Reading, 2021.

[FLE 19] FLEURBAEY M., FERRANNA M., BUDOLFSON M. et al., "The social cost of carbon: Valuing inequality, risk, and population for climate policy", *The Monist*, vol. 102, no. 1, pp. 84–109, Oxford University Press, Oxford, 2019.

[FRA 15] FRANOIS, *Loué sois-tu*, Bayard, Montrouge, 2015.

[FRA 21] FRAUNHOFER-INSTITUT FÜR SOLARE ENERGIESYSTEME ISE, Forschen für die Energiewende, available at: https://www.ise.fraunhofer.de, 2021.

[FOU 66] FOUCAULT M., *Les Mots et les choses. Une archéologie des sciences humaines*, Gallimard, Paris, 1966.

[GIO 14] GIORGINI P., *La transition fulgurante*, Bayard, Montrouge, 2014.

[GIO 16a] GIORGINI P., ARÈNES J., *Au crépuscule des lieux*, Bayard, Montrouge, 2016.

[GIO 16b] GIORGINI P., VAILLANT N., *La fulgurante recréation*, Bayard, Montrouge, 2016.

[GIO 18] GIORGINI P., DEPREZ S., WALD LASOWSKI A., *La tentation d'Eugénie. L'humanité face à son destin*, Bayard, Montrouge, 2018.

[GIO 20] GIORGINI P., *La crise de la joie*, Bayard, Montrouge, 2020.

[GIO 21a] GIORGINI P., MAGNIN T., *Vers une civilisation de l'algorithme ? Un regard chrétien sur un défi éthique*, Bayard, Montrouge, 2021.

[GIO 21b] GIORGINI P., *The Contributory Revolution*, ISTE Ltd, London, and John Wiley & Sons, New York, 2021.

[GIR 14] GIRAUD G., KAHRAMAN Z., How dependent is output growth from primary energy?, available at: https://www.parisschoolofeconomics.eu/IMG/pdf/13juin-pse-ggiraud-presentation-1.pdf, 2014.

[GLO 21] GLOBAL MONITORING LABORATORY, Trends in atmospheric carbon dioxide, available at: https://gml.noaa.gov/ccgg/trends/mlo.html, 2021.

[GOD 15] GODARD O., *La justice climatique mondiale*, La Découverte, Paris, 2015.

[GOL 72] GOLDSMITH E., ALLEN R., ALLABY M. et al., "A blueprint for survival", *The Ecologist*, vol. 2, no. 1, pp. 1–22, 1972.

[GOS 04] GOSSERIES A., *Penser la justice entre les générations. De l'affaire Perruche à la réforme des retraites*, Aubier, Paris, 2004.

[GRA 16] GRANDJEAN A., MARTINI M., *Financer la transition énergétique*, Les Éditions de l'atelier, Ivry-Sur-Seine, 2016.

[GRE 13] GREENPEACE, Scénario de transition énergétique, available at: https://cdn.greenpeace.fr/site/uploads/2017/02/Scenario-Transition-Energetique-Greenpeace-2013.pdf, 2013.

[GRÜ 12] GRÜNE-YANOFF T., "Old wine in new casks: Libertarian paternalism still violates liberal principles", *Social Choice and Welfare*, vol. 38, no. 4, pp. 635–645, 2012.

[HAM 20] HAMMARBY SJÖSTAD 2.0, Electricity, available at: https://hammarbysjostad20.se/?lang=en, 2020.

[HAN 16] HANSEN P.G., "The definition of nudge and libertarian paternalism: Does the hand fit the glove?", *European Journal of Risk Regulation*, vol. 1, p. 155, 2016.

[HAU 10] HAUSMAN D.M., WELCH B., "Debate: To nudge or not to nudge", *Journal of Political Philosophy*, vol. 18, no. 1, pp. 123–136, 2010.

[HEY 16] HEYWARD C., ROSER D., *Climate Justice in a Non-Ideal World*, Oxford University Press, New York, 2016.

[HOE 18] HOEGH-GULDBERG O., JACOB D., BINDI M. et al., Impacts of 1.5 °C global warming on natural and human systems, Report, IPCC Secretariat, 2018.

[INS 22] INSEE, Émissions de gaz à effet de serre par activité, available at: https:///www.insee.fr/fr/statistiques/2015759, 2022.

[INT 18] INTERGOVERNMENTAL PANEL ON CLIMATE CHANGE, Global warming of 1.5°C, Summary for Policymakers, Report, IPCC, Geneva, 2018.

[IPC 01] IPCC TAR, Climate change 2001: Synthesis report, IPCC, Geneva, 2001.

[JAM 02] JAMIESON D., *Morality's Progress: Essays on Humans, Other Animals, and the Rest of Nature*, Oxford University Press, Oxford, 2002.

[JAN 17] JANETOS A.C., *Recommendations of the Climate Action Task Force for Boston University's Climate Action Plan*, Boston University, Boston, 2017.

[JAN 20a] JANCOVICI J.M., Qu'est-ce que l'équation de Kaya ?, available at: https://jancovici.com/changement-climatique/economie/quest-ce-que-lequation-de-kaya/, 2020.

[JAN 20b] JANCOVICI J.M., Combien de gaz à effet de serre dans notre poubelle ?, available at: https://jancovici.com/changement-climatique/les-ges-et-nous/combien-de-gaz-a-effet-de-serre-dans-notre-poubelle, 2020.

[JEV 66] JEVONS W.S., *The Coal Question: An Inquiry Concerning the Progress of the Nation and the Probable Exhaustion of Our Coal Mines*, Macmillan & Co, London, 1866.

[KOE 07] KOEPPEL, S., ÜRGE-VORSATZ, D., Assessment of policy instruments for reducing greenhouse gas emissions from buildings, A Report of UNEP Sustainable Buildings and Construction Initiative, United Nations Environmental Program and Central European University, Paris, 2007.

[LAB 15] LABBOUZ-HENRY D., Bâtiments tertiaires performants et comportements favorables à l'environnement : le rôle de variables psychosociales et du contexte organisationnel, Social psychology PhD Thesis, Université Paris-Nanterre, Nanterre, 2015.

[LAT 91] LATOUR B., CALLON M. (eds), *La science telle qu'elle se fait. Anthologie de la sociologie des sciences de langue anglaise*, La Découverte, Paris, 1991.

[LAV 18] LAVERGNE R., "Climat et transition énergétique", *Annales des Mines – Responsabilité et environnement*, vol. 89, no. 1, pp. 39–43, 2018.

[LEG 15] LE GRAND J., NEW B., *Government Paternalism – Nanny State or Helpful Friend*, Princeton University Press, Princeton, 2015.

[LÉV 14] LÉVY J.P., ROUDIL N., FLAMAND A. et al., "Les déterminants de la consommation énergétique domestique", *Flux*, no. 96, pp. 40–54, 2014.

[LIG 96] LIGHT A., KATZ E., "Introduction: Environmental pragmatism and environmental ethics as contested terrain", in LIGHT A., KATZ E. (eds.), *Environmental Pragmatism*, Taylor & Francis, London, 1996.

[LLA 15] LLAVADOR H., ROEMER J.E., SYLVESTRE J., *Sustainability for a Warming Planet*, Harvard University Press, Cambridge, 2015.

[MAN 20] MANCHESTER 1824, The University of Manchester: A review of ES performance, available at: https://documents.manchester.ac.uk/DocuInfo.aspx?DocID=46576, 2020.

[MAR 07] MARKANDYA A., WILKINSON P., "Electricity generation and health", *The Lancet*, vol. 370, pp. 979–990, 2007.

[MCC 20] MCCORNASKY O., Copenhagen's Zero Carbon Plan, available at: https://zerocarbon2025.com/copenhagens-zero-carbon-plan/, 2020.

[MCK 17] MCKINNON C., "Endangering humanity: An international crime?", *Canadian Journal of Philosophy*, vol. 47, nos 2–3, pp. 395–415, 2017.

[MCM 81] MCMAHAN J., "Problems of population theory", *Ethics*, vol. 92, no. 1, pp. 96–127, 1981.

[MEM 16] MEMOORI, Smart Buildingx cqn be "The Nodes" of the smart grid, www.memoori.com/smart-buildings-can-nodes-smart-grid, 2016.

[MÉR 21] MÉRITET S., "On pourra lire à ce propos l'article très éclairant paru dans le journal", *Le Monde*, available at: https://www.lemonde.fr/idees/article/2021/03/17/panne-electrique-au-texas-c-est-la-complementarite-et-non-la-substitution-entre-marches-et-regulations-qu-il-faut-mettre-en-place_6073401_3232.html, 2021.

[MÉT 17] MÉTHOS, De l'usage des bâtiments performants en Région Bruxelles Capitale, Report, Méthos, Brussels, 2017.

[MEY 12] MEYER L.H., ROSEN D., "Enough for the future", in MEYER L.H. (ed.), *Intergenerational Justice*, Routledge, London, 2012.

[MEY 21] MEYER L., Intergenerational Justice, available at: https://plato.stanford.edu/archives/sum2021/entries/justice-intergenerational/, 2021.

[MIC 03] MICHELETTI M., *Political Virtue and Shopping: Individuals, Consumerism, and Collective Action*, Palgrave Macmillan, New York, 2003.

[MIN 21] MINISTÈRE DE LA TRANSITION ÉCOLOGIQUE ET DE LA COHÉSION DES TERRITOIRES, MINISTÈRE DE LA TRANSITION ÉNERGÉTIQUE, "Fit for 55" : un nouveau cycle de politiques européennes pour le climat, available at: https://www.ecologie.gouv.fr/fit-55-nouveau-cycle-politiques-europeennes-climat, 2021.

[MIT 05] MITCHELL G., "Libertarian paternalism is an oxymoron", *Northwestern University Law Review*, vol. 99, no. 3, pp. 1245–1277, 2005.

[MOR 11] MORTON J.M, "Toward an ecological theory of the norms of practical deliberation", *European Journal of Philosophy*, vol. 19, no. 4, pp. 561–584, 2011.

[MOU 11] MOUHAUT J.F., *Des esclaves énergétiques*, Champ Vallon, Ceyzérieu, 2011.

[MRM 21] MR MONDIALISATION, La Chine et son programme sans précédent de modification météo, available at: https://mrmondialisation.org/la-chine-et-son-programme-sans-precedent-de-modification-meteo/, 2021.

[MUL 11] MULGAN T., *Ethics for a Broken World: Imagining Philosophy After Catastrophe*, Routledge, Durham, 2011.

[MUL 15] MULGAN T., "Utilitarianism for a broken world", *Utilitas*, vol. 27, no. 1, pp. 92–114, 2015.

[NAT 21] NATURE COMMUNICATIONS, Altered growth conditions more than reforestation counteracted forest biomass carbon emissions 1990–2020, available at: https://www.nature.com/articles/s41467-021-26398-2, 2021.

[NEC 15] NEC The Global Commission on the Economy and Climate, *La nouvelle économie climatique. Une meilleure croissance, un meilleur climat*, Les Petits Matins, Paris, 2015.

[NÉG 21] NÉGAWATT, La transition énergétique au cœur d'une transition sociétale, https://negawatt.org/Scenario-negaWatt-2022, 2021.

[NOR 94] NORTON B.G., *Toward Unity among Environmentalists*, Oxford University Press, Oxford, 1994.

[OGI 04] OGIEN R., *La panique morale*, Grasset, Paris, 2004.

[ONE 92] O'NEILL J., "The varieties of intrinsic value", *The Monist*, vol. 75, no. 2, pp. 119–137, 1992.

[ORS 18] ORSENNA E., QUÉRÉ S., *La Fabrique du neuf*, Le Cherche midi, Paris, 2018.

[PAG 07] PAGE E.A., *Climate Change, Justice and Future Generations*, Edward Elgar Publishing, London, 2007.

[PAS 74] PASSMORE J.A., *Man's Responsibility for Nature: Ecological Problems and Western Traditions*, Duckworth, London, 1974.

[RAB 01] RABL A., SPADARO J.V., "Les coûts externes de l'électricité", *Revue de l'énergie*, no. 525, pp. 151–163, 2001.

[RAW 09] RAWLS J., *Théorie de la justice*, translated by AUDARD C., Points, Paris, 2009.

[RAZ 86] RAZ J., *The Morality of Freedom*, Clarendon Press, Oxford, 1986.

[RIF 12] RIFKIN J., *La Troisième Révolution Industrielle*, Les Liens qui libèrent, Paris, 2012.

[ROB forthcoming] ROBYNS B. et al. *Smart Grids and Buildings for Energy and Societal Transition*, ISTE Ltd, London and John Wiley & Sons, New York, forthcoming.

[ROB 15] ROBYNS B., FRANÇOIS B., DELILLE G. et al., *Energy Storage in Electric Power Grids*, ISTE Ltd, London, and John Wiley & Sons, New York, 2015.

[ROB 16] ROBYNS B., SAUDEMONT C., HISSEL D. et al., *Electrical Energy Storage in Transportation Systems*, ISTE Ltd, London, and John Wiley & Sons, New York, 2016.

[ROB 19] ROBYNS B., DAVIGNY A., BARRY H. et al., *Electrical Energy Storage for Buildings in Smart Grids*, ISTE Ltd, London, and John Wiley & Sons, New York, 2019.

[ROB 21] ROBYNS B., DAVIGNY A., FRANÇOIS B. et al., *Electric Power Generation from Renewable Sources*, 2nd edition, ISTE Ltd, London, and John Wiley & Sons, New York, 2021.

[ROS 02] ROSS D., *The Right and the Good*, Oxford University Press, Oxford, 2002.

[ROS 16] ROSER D., SEIDEL C., *Climate Justice: An Introduction*, Routledge, Abingdon, 2016.

[ROZ 20] ROZEBOOM G.J., "Nudging for rationality and self-governance", *Ethics*, vol. 131, no. 1, pp. 107–121, 2020.

[RTE 21] RTE, Futurs énergétiques 2050, Report, RTE, 2021.

[SCH 13] SCHEFFLER S., *Death and the Afterlife*, Oxford University Press, Oxford, 2013.

[SCH 19] SCHMIDT A.T., "Getting real on rationality – Behavioral science, nudging, and public policy", *Ethics*, vol. 129, no. 4, pp. 511–543, 2019.

[SÉN 20] SÉNAT FRANÇAIS, Pour une transition numérique écologique, Report, Sénat, 2020.

[SHA 21] SHAHROUR I., *Voyage au cœur de la ville intelligente*, Édition Isam Shahrour, 2021.

[SHU 88] SHUE H., "Mediating duties", *Ethics*, vol. 98, no. 4, pp. 687–704, 1988.

[SHU 14] SHUE H., *Climate Justice: Vulnerability and Protection*, Oxford University Press, Oxford, 2014.

[SIN 15] SINGLER E., *Green nudge : réussir à changer les comportements pour sauver la planète*, Pearson Education, Montreuil, 2015.

[STE 20] STEPHANT M., ABBES D., HASSAM-OUARI K. et al., "Increasing photovoltaic self-consumption with game theory and blockchain", *EAI Endorsed Transactions on Energy Web*, vol. 8, no. 34, pp. 1–12, 2020.

[STE 21] STEPHANT M., ABBES D., HASSAM-OUARI K. et al., "Distributed optimization of energy profiles to improve photovoltaic self-consumption on a local energy community", *Simulation Modelling Practice and Theory*, vol. 108, pp. 1–13, 2021.

[STO 22] STOCKHOLM UNIVERSITY, Stockholm University Climate Roadmap for the period 2020-2040, available at: https://www.su.se/staff/organisation-governance/governing-documents-rules-and-regulations/environment/stockholm-university-climate-roadmap-for-the-period-2020-2040-1.536900, 2022.

[SUB 09] SUBRÉMON H., Habiter avec l'énergie, pour une anthropologie sensible de la consommation énergétique, PhD thesis, Université Paris-Nanterre, 2009.

[SUB 15] SUBRÉMON H., AUBRY-BRÉCHAIRE M., BOUSQUET P. et al., "Coriolis à l'usage : un bâtiment performant sous tension", *Métropolitiques*, available at: http://www.metropolitiques.eu/Coriolis-a-l-usage-un-batiment.html, 2015.

[SUN 03] SUNSTEIN C.R., THALER R.H., "Libertarian paternalism is not an oxymoron", *The University of Chicago Law Review*, vol. 70, no. 4, pp. 1159–1202, 2003.

[SUN 06] SUNSTEIN C.R., "Preferences, paternalism, and liberty", *Royal Institute of Philosophy Supplements*, vol. 59, pp. 233–264, 2006.

[SUN 08] SUNSTEIN C.R., THALER R.H., *Nudge: Improving Decisions about Health, Wealth, and Happiness*, Yale University Press, London, 2008.

[SUN 14] SUNSTEIN C.R., "Nudging: A very short guide", *Journal of Consumer Policy*, vol. 37, no. 4, pp. 583–588, 2014.

[SUN 15] SUNSTEIN C.R., *Why Nudge? The Politics of Libertarian Paternalism*, Yale University Press, Yale, 2015.

[THA 15] THALER R.H., *Misbehaving: The Making of Behavioral Economics*, W.W. Norton & Company, New York, 2015.

[THE 19] THE GUARDIAN, Inside Copenhagen's race to be the first carbon-neutral city, available at: https://www.theguardian.com/cities/2019/oct/11/inside-copenhagens-race-to-be-the-first-carbon-neutral-city, 2019.

[THØ 09] THØGERSEN J., CROMPTON T., "Simple and painless? The limitations of spillover in environmental campaigning", *Journal of Consumer Policy*, vol. 32, no. 2, pp. 141–163, 2009.

[UNI 22] UNIVERSITY OF BRITISH COLUMBIA, UBC Vancouver Climate Action Plan 2030, University of British Colombia, 2022.

[UNE 21] UN ENVIRONMENT PROGRAMME, Over 1,000 universities and make net-zero pledges as new nature initiative is unveiled, available at: https://www.unep.org/news-and-stories/press-release/over-1000-universities-and-colleges-make-net-zero-pledges-new-nature, 2021.

[UNI 19a] UNIVERSITÉ CATHOLIQUE DE LILLE, "Ville vs nature ? La biodiversité des initiatives de l'université", *Live TREE Mag*, no. 5, pp. 12–21, 2019a.

[UNI 19b] UNIVERSITÉ CATHOLIQUE DE LILLE, "Associations étudiantes", *Live TEE Mag*, no. 5, p. 24, June 2019.

[VIL 18] VILLABLA B., SÈMAL L., *Sobriété énergétique : contraintes matérielles, équité sociale et perspectives institutionnelles*, Quae, Paris, 2018.

[WAR 20] WARNOCK G.J., *The Object of Morality*, Routledge, London, 2020.

[WHI 13] WHITE M.D., *The Manipulation of Choice: Ethics and Libertarian Paternalism*, Palgrave Macmillan, New York, 2013.

[WHI 21] WHITFORD E., Global Universities announce carbon neutrality alliance, available at: https://www.insidehighered.com/quicktakes/2021/11/02/global-universities-announce-carbon-neutrality-alliance, 2021.

[WIK 20a] WIKIPEDIA, Électricité en Europe, available at: https://fr.wikipedia.org/wiki/%C3%89lectricit%C3%A9_en_Europe, 2020.

[WIK 20b] WIKIPEDIA, Sobriété énergétique et effet de serre, 2020.

[WIK 20c] WIKIPEDIA, Énergie et effet de serre, available at: https://fr.wikipedia.org/wiki/%C3%89nergie_et_effet_de_serre, 2020.

[WIK 20d] WIKIPEDIA, Recherche-action, available at: https://fr.wikipedia.org/wiki/Recherche-action, 2020.

[WIL 13] WILKINSON T.M., "Nudging and manipulation", *Political Studies*, vol. 61, no. 2, pp. 341–355, 2013.

[WWF 16] WWF/HEAL/CAN EUROPE/SANDBAG, Europe's Dark Cloud: How coal-burning countries are making their neighbours sick, Report, WWF/HEAL/CAN Europe/Sandbag, 2016.

Index

A, B

C

D, E

F, G, I

L, M, N

O, P

Q, R

S

T, U, V

Other titles from

in

Energy

2022

ABOUELATTA Mohamed, SHAKER Ahmed, GONTRAND Christian
Smart Power Integration

BOUTAUD Benoit
Energy Autonomy: From the Notion to the Concepts

2021

ROBYNS Benoît, DAVIGNY Arnaud, FRANÇOIS Bruno, HENNETON Antoine, SPROOTEN Jonathan
Electricity Production from Renewable Energies (2nd edition)

2020

BOISGIBAULT Louis, AL KABBANI Fahad
Energy Transition in Metropolises, Rural Areas and Deserts

SOUALHI Abdenour, RAZIK Hubert
Electrical Systems 1: From Diagnosis to Prognosis
Electrical Systems 2: From Diagnosis to Prognosis

2019

BENALLOU Abdelhanine
Energy Transfers by Convection
(Energy Engineering Set – Volume 3)
Energy Transfers by Radiation
(Energy Engineering Set – Volume 4)
Mass Transfers and Physical Data Estimation
(Energy Engineering Set – Volume 5)

LACHAL Bernard
Energy Transition

ROBYNS Benoît, DAVIGNY Arnaud, BARRY Hervé, KAZMIERCZAK Sabine, SAUDEMONT Christophe, ABBES Dhaker, FRANÇOIS Bruno
Electrical Energy Storage for Buildings in Smart Grids

2018

BENALLOU Abdelhanine
Energy and Mass Transfers: Balance Sheet Approach and Basic Concepts
(Energy Engineering Set – Volume 1)
Energy Transfers by Conduction
(Energy Engineering Set – Volume 2)

JEMEÏ Samir
Hybridization, Diagnostic and Prognostic of Proton Exchange Membrane Fuel Cells: Durability and Reliability

RUFFINE Livio, BROSETA Daniel, DESMEDT Arnaud
Gas Hydrates 2: Geoscience Issues and Potential Industrial Applications

VIALLET Virginie, FLEUTOT Benoit
Inorganic Massive Batteries
(Energy Storage – Batteries, Supercapacitors Set – Volume 4)

2017

BROSETA Daniel, RUFFINE Livio, DESMEDT Arnaud
Gas Hydrates 1: Fundamentals, Characterization and Modeling

LA SCALA Massimo
From Smart Grids to Smart Cities: New Challenges in Optimizing Energy Grids

MOLINA Géraldine, MUSY Marjorie, LEFRANC Margot
Building Professionals Facing the Energy Efficiency Challenge

SIMON Patrice, BROUSSE Thierry, FAVIER Frédéric
Supercapacitors Based on Carbon or Pseudocapacitive Materials
(Energy Storage – Batteries, Supercapacitors Set – Volume 3)

2016

ALLARD Bruno
Power Systems-on-Chip: Practical Aspects of Design

ANDRÉ Michel, SAMARAS Zissis
Energy and Environment

DUFOUR Anthony
Thermochemical Conversion of Biomass for the Production of Energy and Chemicals

2015

CROGUENNEC Laurence, MONCONDUIT Laure, DEDRYVÈRE Rémi
Electrodes for Li-ion Batteries
(Energy Storage – Batteries, Supercapacitors Set – Volume 2)

LEPRINCE-WANG Yamin
Piezoelectric ZnO Nanostructure for Energy Harvesting
(Nanotechnologies for Energy Recovery Set – Volume 1)

ROBYNS Benoît, FRANÇOIS Bruno, DELILLE Gauthier, SAUDEMONT Christophe
Energy Storage in Electric Power Grids

ROSSI Carole
Al-based Energetic Nanomaterials
(Nanotechnologies for Energy Recovery Set – Volume 2)

TARASCON Jean-Marie, SIMON Patrice
Electrochemical Energy Storage
(Energy Storage – Batteries, Supercapacitors Set – Volume 1)

2013

LALOUI Lyesse, DI DONNA Alice
Energy Geostructures: Innovation in Underground Engineering

2012

BECKERS Benoit
Solar Energy at Urban Scale

ROBYNS Benoît, DAVIGNY Arnaud, FRANÇOIS Bruno, HENNETON Antoine, SPROOTEN Jonathan
Electricity Production from Renewable Energies

2011

GAO Fei, BLUNIER Benjamin, MIRAOUI Abdellatif
Proton Exchange Membrane Fuel Cell Modeling

MULTON Bernard
Marine Renewable Energy Handbook

2010

BRUNET Yves
Energy Storage

2009

SABONNADIÈRE Jean-Claude
Low Emission Power Generation Technologies and Energy Management

SABONNADIÈRE Jean-Claude
Renewable Energy Technologies

Printed and bound by CPI Group (UK) Ltd, Croydon, CR0 4YY

10/09/2023

08112074-0001